T889

Problem solving and improvement: quality and other approaches

The Open University

Block 3 Techniques

This publication forms part of an Open University course T889 *Problem solving and improvement: quality and other approaches*. Details of this and other Open University courses can be obtained from the Student Registration and Enquiry Service, The Open University, PO Box 197, Milton Keynes MK7 6BJ, United Kingdom: tel. +44 (0)845 300 60 90, email general-enquiries@open.ac.uk

Alternatively, you may visit the Open University website at http://www.open.ac.uk where you can learn more about the wide range of courses and packs offered at all levels by The Open University.

To purchase a selection of Open University course materials visit http://www.ouw.co.uk, or contact Open University Worldwide, Michael Young Building, Walton Hall, Milton Keynes MK7 6AA, United Kingdom for a brochure. tel. +44 (0)1908 858793; fax +44 (0)1908 858787; email ouw-customer-services@open.ac.uk

The Open University
Walton Hall, Milton Keynes
MK7 6AA

First published 2007.

Edited and designed by The Open University.

Typeset by SR Nova Pvt. Ltd, Bangalore, India.

Printed and bound in the United Kingdom by Hobbs The Printers Limited, Brunel Road, Totton, Hampshire SO40 3WX.

ISBN 978 0 7492 2351 9

1.1

Block 3 Techniques

CONTENTS

AIMS

The aims of Block 3 are to:

- introduce you to a wide range of techniques that can be used to investigate situations, conduct analyses, solve problems and/or devise improvements, implement change, and monitor future performance
- enable you to develop sufficient level of competence to start using these techniques in real-life situations where problem solving and improvement are required.

LEARNING OUTCOMES

After studying Block 3 you should be able to:

- select an appropriate technique to achieve a particular purpose in a given situation
- apply a wide range of techniques to investigate situations, conduct analyses, solve problems and/or devise improvements, implement change and monitor future performance
- present the results of your applications in ways that are appropriate to the techniques being used.

1 INTRODUCTION

At the end of Block 1 I introduced you to three generic problem-solving and improvement methods based on three different metaphors for problem solving:

1 a learning cycle
2 a journey
3 a search.

In Block 4 I shall look at a number of more specialised and more sophisticated methods but in this block I concentrate on the techniques that can be used within these various methods in order to investigate, analyse, generate and implement recommendations, and monitor future performance. There are two main reasons for my decision to put the Techniques block before the Methods block. The first is that it is difficult to explain the stages of some of the methods without reference to the techniques they use. The second is that you can gain substantial benefit from using techniques in the context of the generic methods you saw at the end of Block 1 or even using individual techniques in isolation.

Another point to recognise is that there is no clear dividing line between techniques and methods. Everything in Blocks 3 and 4 can be positioned along the continuum shown in Figure 3.1 but for many of the items there is room for debate over their precise positions. For example, Taguchi methods are clearly labelled as methods, but I have included them under the label of techniques. Similarly, when you come to Block 4 you will find little to distinguish one thing that is given the name methodology (soft systems methodology) from something very similar that is called a method (the hard systems method).

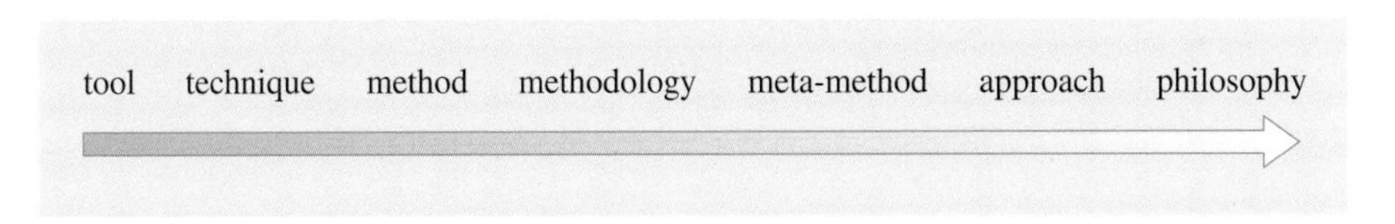

Figure 3.1 From tool to philosophy

There are very many ways in which the techniques covered in this block could have been arranged. I have chosen one that I hope you will find meaningful and useful. It is a combination of:

- very well-known groupings such as the seven 'old' tools
- techniques of a similar type such as questioning techniques and systems diagrams
- techniques that perform a similar function such as idea generation techniques.

It is quite common to classify techniques purely according to the stage in a problem-solving or improvement method in which they are used but I have not done this because many techniques can be used at more than one stage. For example, statistical process control techniques can be used to check the capability of processes, monitor their performance and alert people to a switch from 'in control' to 'out of control', identify opportunities for improvement, and monitor the effectiveness of changes.

The contents list shows you the way the techniques are arranged. As you study each technique you might like to consider where you could use it within the DRIVE model for continuous improvement, and enter it in one or more columns of Table 3.1. (See figure 3.2 for a diagrammatic representation of Oakland and Marosszeky's DRIVE model that you met in Block 1.) At the end of the block you will find my version, but do bear in mind it is not a definitive answer. The subjectivity of much problem-solving and improvement activity applies here too.

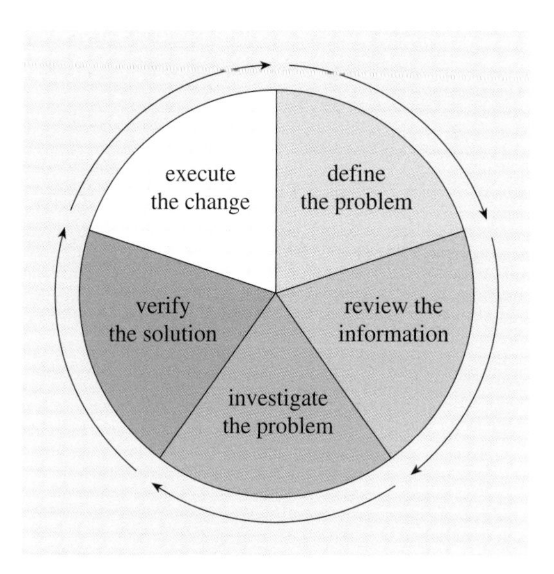

Figure 3.2 A diagrammatic version of the DRIVE model

Table 3.1 Techniques for defining, reviewing, investigating, verifying and executing

Technique	Define	Review	Investigate	Verify	Execute
Pareto analysis					
Cause-and-effect diagrams					
Stratification					
Tally cards					
Histograms					
Scatter diagrams					
Shewhart control charts					
Affinity diagrams					
Relational diagrams					
Tree diagrams					
Matrix diagrams					
Program decision process charts					
Arrow diagrams					
Matrix data analysis					
Is/is not analysis					
Five whys					
Five Ws and H					
More specific sets of questions					
Input-output diagrams					
Systems maps					
Influence diagrams					
Rich pictures					
Activity sequence diagrams					
Flow-block diagrams					
Flow-process diagrams					
Spaghetti charts					
SIPOC charts					
Multiple-cause diagrams					
Force field analysis					
Cognitive mapping					

Technique	Define	Review	Investigate	Verify	Execute
Failure modes and effects analysis					
Fault tree analysis					
Brainstorming					
Brainwriting					
Nominal group technique					
SCAMPER					
Creative problem solving					
Six hats and lateral thinking					
Stakeholder analysis					
SWOT					
Environmental scanning					
Benchmarking					
Gap analysis					
Poka-yoke					
Quality function deployment					
TRIZ					
Y2X					
Taguchi methods					
SMED					
Process redesign					
Process capability analysis					
Control charts					
Weighted score method					
Decision trees					
Solution selection matrix					
Risk assessment					
A simple step procedure for implementation					
Gantt charts					
Key performance indicators					
Radar diagrams					
Balanced scorecard					

2 THE SEVEN 'OLD' TOOLS

The data-gathering and analytical work of quality circles and similar groups has traditionally drawn on a set of simple, but effective, statistical methods. These methods became known as the seven tools, but have been renamed as the seven 'old' tools (also known as the seven quality control tools or the seven basic tools) to distinguish them from another collection of methods known as the seven 'new' tools. The 'old' tools are:

- Pareto analysis
- cause-and-effect diagrams
- stratification
- tally cards
- histograms
- scatter diagrams
- Shewhart control charts.

You met five of the seven 'old' tools in Block 2, so only brief reminders of those are given here.

2.1 Pareto analysis

An Italian economist called Vilfredo Pareto has been credited with the development of this histogram-based technique that you met in Section 2.4 of Block 2. As you may recall, it is often used in deciding where to focus improvement activities, and the craft knowledge that lies behind it is the frequently observed phenomenon that in many situations something like 80% of problems are usually attributable to just a 'vital few' sources. Pareto, often referred to as the 80:20 rule, can help you to identify these 'vital few' sources.

2.2 Cause-and-effect diagrams

Cause-and-effect diagrams are often known as fishbone diagrams because of their appearance, or Ishikawa diagrams after Kaoru Ishikawa who pioneered the quality circle movement in Japan in the 1950s. A cause-and-effect diagram is constructed by starting with the effect (the 'backbone') and then building up a network of causes and sub-causes (the 'ribs') as shown in Figure 3.3. Oakland (1986) recommends the use of Lockyer's five 'P's of production management (Product, including materials and intermediates; Processes or methods of manufacture; Plant; Programmes or timetables for ordering, manufacture and shipment; People) as the main 'ribs' of the diagram.

Figure 3.4 shows a completed diagram of the causes of hospital beds not being available for emergency admissions.

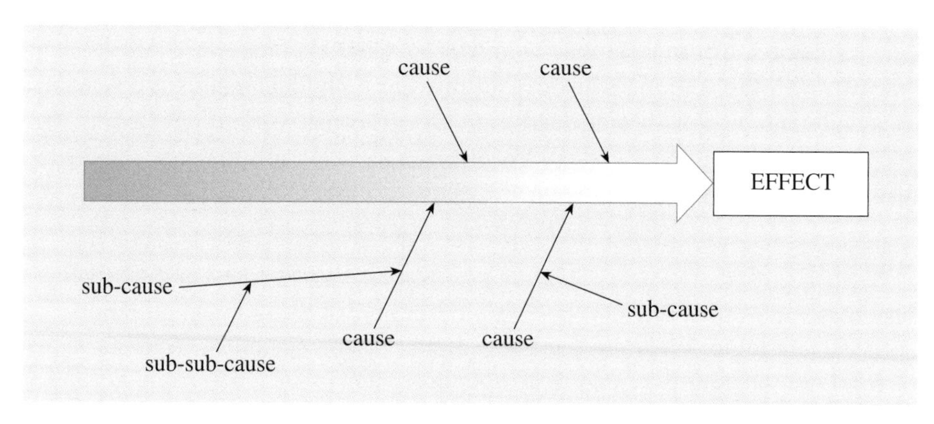

Figure 3.3 General form of a cause-and-effect diagram

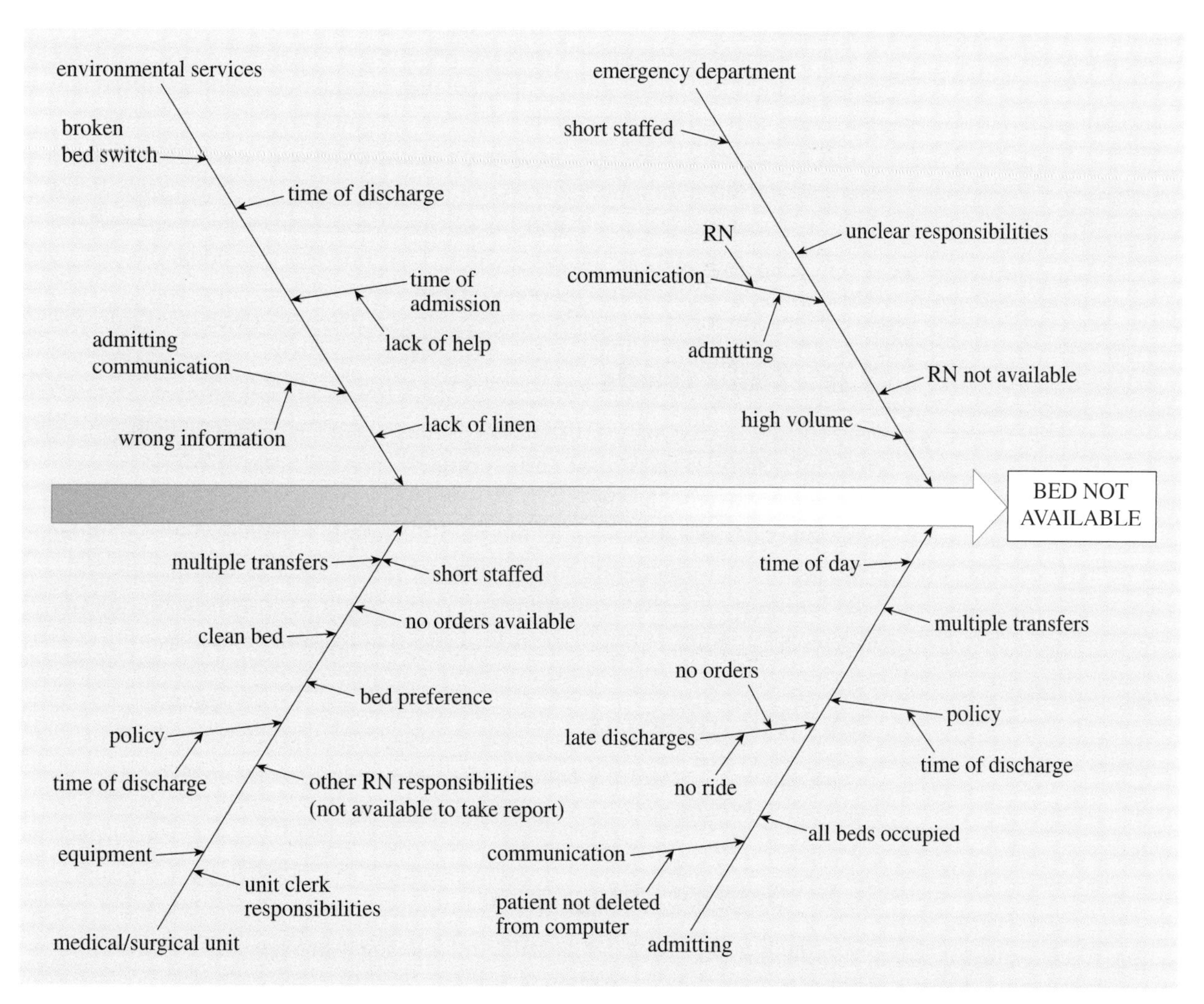

Figure 3.4 Cause-and-effect diagram: causes of beds not being available (Source: Evans and Lindsay, 1996, p. 369)

One refinement of the technique is known as CEDAC (cause-and-effect diagram with the addition of cards). In this the individual causes are written on small cards or Post-it notes and a group of people work at developing the diagram by moving the cards or Post-it notes around until they are satisfied that they are in the correct places.

ACTIVITY 3.1 ....

The cause-and-effect diagram has been described as using systematic but not systemic logic. What are the implications of this statement in terms of what the diagram does and does not show? ●

2.3 Stratification

You met the term 'stratification' in Section 2.5 of Block 2. It is the division of data into classes so that subsets can be studied separately. These different subsets might be data from different sources, or data collected at different times or under different conditions, and so on. Stratified data may make trends or abnormalities more noticeable than they would have been if the data had not been stratified. For example, if the same part is being made on three different machines, an analysis of the dimensions of the parts made on each machine may be more revealing than one that considers the dimensions of all the parts produced irrespective of their source.

2.4 Tally cards

As you saw in Section 2.1 of Block 2, a tally chart is a chart which is used to keep a 'tally' or record of certain objects, observations or occurrences. Tally cards are just the means of recording the tallies. In addition to recording attribute data such as the on/off example in Block 2, they can be used to record variations in quantifiable attributes such as dimensions and thus build up a frequency distribution. An example is shown in Figure 3.5. It depicts variation in the level of lead in drinking water samples.

2.5 Histograms

A histogram is a form of frequency distribution diagram. I introduced histograms in Section 2.2 of Block 2.

2.6 Scatter diagrams

These help you to look for possible relationships between two variables. You looked at them in Section 2.5 of Block 2.

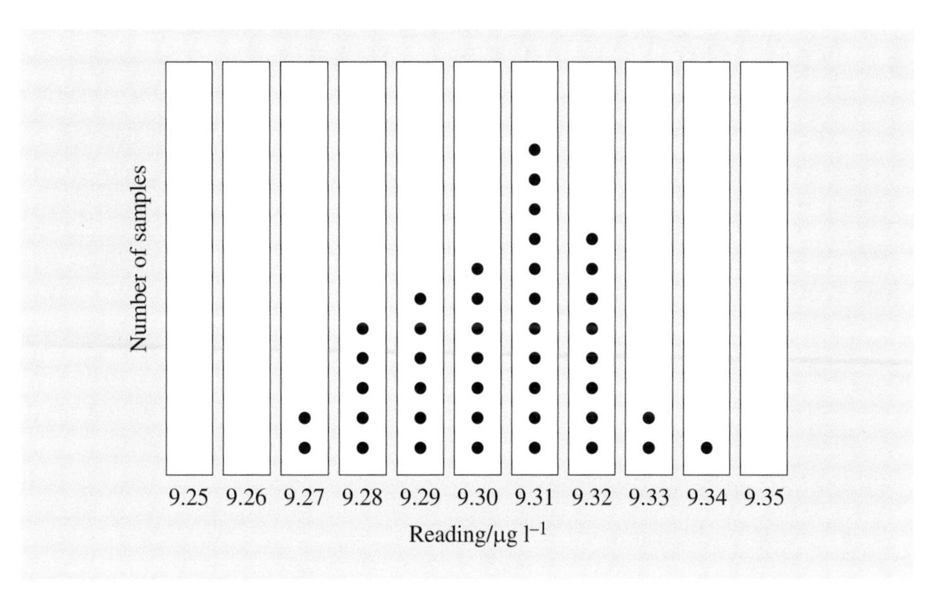

Figure 3.5 Tally card showing variation in the level of lead in drinking water samples

2.7 Shewhart control charts

Shewhart control charts are one of the main tools used in statistical process control. They will be considered in depth in Section 11 of this block.

3 THE SEVEN 'NEW' TOOLS

Use of the seven 'new' tools began in Japan in the 1970s but they did not become widely known elsewhere until more than a decade later. A less common name for them is the management and planning (MP) tools.

Cohen (1988) draws the following distinction between the 'old' and the 'new' tools:

> [The old] tools help people to analyze data pertaining to events that have already occurred and to understand the processes that helped to shape those events. These tools are especially useful for data analysis, cause-and-effect analysis, and process management.
>
> The 'Seven New Tools' are:
>
> - Affinity diagram;
> - Relational diagram;
> - Tree diagram;
> - Matrix diagram;
> - Program decision process chart;
> - Arrow diagram; and
> - Matrix data analysis.
>
> These tools help us to make decisions about the future. They help people to analyze relationships between ideas and attributes – intangibles that are difficult or impossible to quantify. They help us to structure ideas and map their interrelationships.
>
> (Cohen, 1988, pp. 198–9)

3.1 Affinity diagrams

An affinity diagram is a tool for organising non-numerical information into groupings based on natural relationships between items of information. One way of building up the diagram is as follows.

1. The individual pieces of information or ideas are each written on a Post-it note.
2. The Post-it notes are displayed.
3. Notes containing ideas that appear to be related are sorted into groups.
4. Headings that capture the meaning/spirit of each group are devised.
5. The various groups are organised into a single diagram.

An example of an affinity diagram is shown in Figure 3.6.

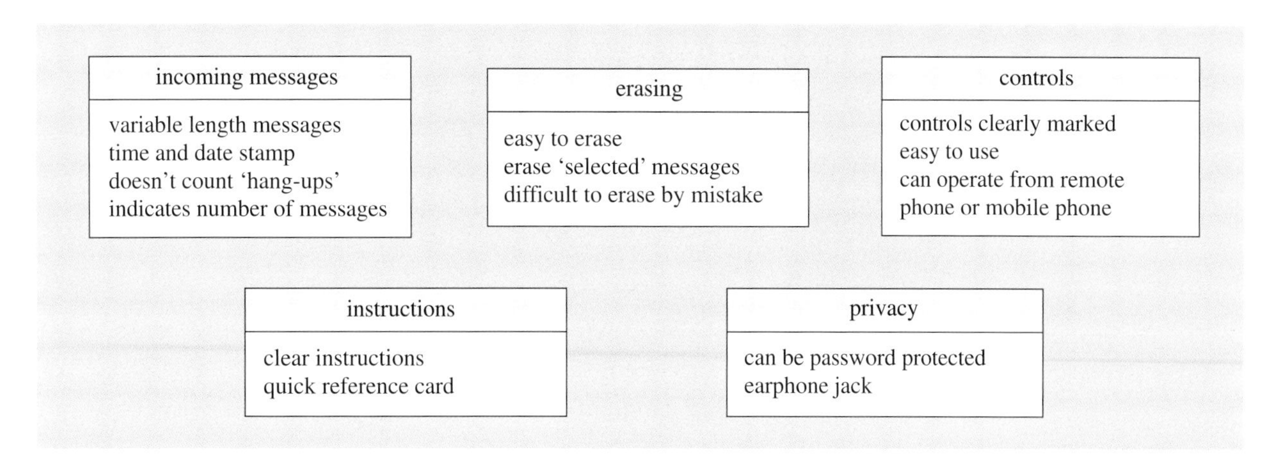

Figure 3.6 Affinity diagram for a telephone answering machine

ACTIVITY 3.2

I have conducted an exercise to see how the number of typographic errors in this block can be reduced. I have asked the team to suggest factors that might be important to consider. Their suggestions are shown in the list below. Arrange them into an affinity diagram and then compare your attempt with mine, which is given at the end of the block.

editing skill
grammar
interruptions during
preparation
font size
print quality
keyboard skill
desk height
technical terms
noise
spell check
handwriting of original
changes to drafts
font type
grammar check
interruptions during
proofreading
colour of print
temperature
keyboard
punctuation
time of day
lighting
proofreading skill
changes to final
version
position of screen
no feedback
tight deadlines
spelling
chair height
foreign words
auto correction

3.2 Relational diagrams

Relational diagrams show the structure of relationships. Figure 3.7 shows the relationships between the various departments in a manufacturing company that is experiencing difficulties in introducing new products into production. The ovals contain components and the lines joining them mean 'related to' or 'affects'.

3.3 Tree diagrams

A tree diagram is a way of arriving at a hierarchically structured set of ideas and can also be used to map out the tasks needed to fulfil a primary goal via sub-goals. It is often the next stage of analysis after an affinity diagram has been drawn; analytical skills are used to fill in the gaps at each level of the

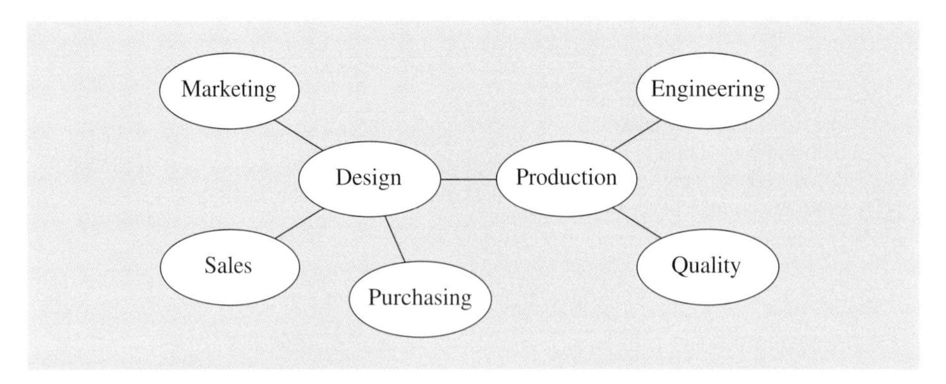

Figure 3.7 Relational diagram

hierarchy. Figure 3.8 shows a tree diagram based on the affinity diagram shown in Figure 3.6.

3.4 Matrix diagrams

A matrix diagram enables ideas in one dimension, say problem characteristics, to be mapped against ideas in another dimension, say solution criteria. One set of ideas is arranged as the columns of a matrix and the other as the rows. The relationship between each pair of ideas is then examined using a predetermined procedure and the result of each comparison is recorded in the appropriate cell of the matrix. The diagram thus allows a cumulative pattern to be built up from the results of the interactions between the pairs. In Figure 3.9 a matrix diagram has been used to investigate delivery problems.

3.5 Program decision process charts

These charts, which it is safer to call by their initials PDPC because many users refer to them as process decision program charts, are concerned with the activities and decisions involved in transforming inputs into outputs. They can be used to investigate opportunities for improvement by developing a more detailed understanding of how a process works, and can also allow potential problem areas and bottlenecks to be predicted by examining how various steps in the process relate to each other.

Unfortunately, there is no consensus about what a PDPC should look like or precisely what information it should convey. I have included a couple of the many variations. Figure 3.10 shows an interpretation that emphasises decisions but has very little to say about process. Figure 3.11 looks at various elements of a programme to detect potential problems and identify potential countermeasures but does not seem much concerned with decisions unless you count the indications shown by O and X of which suggestions to take up and which to reject.

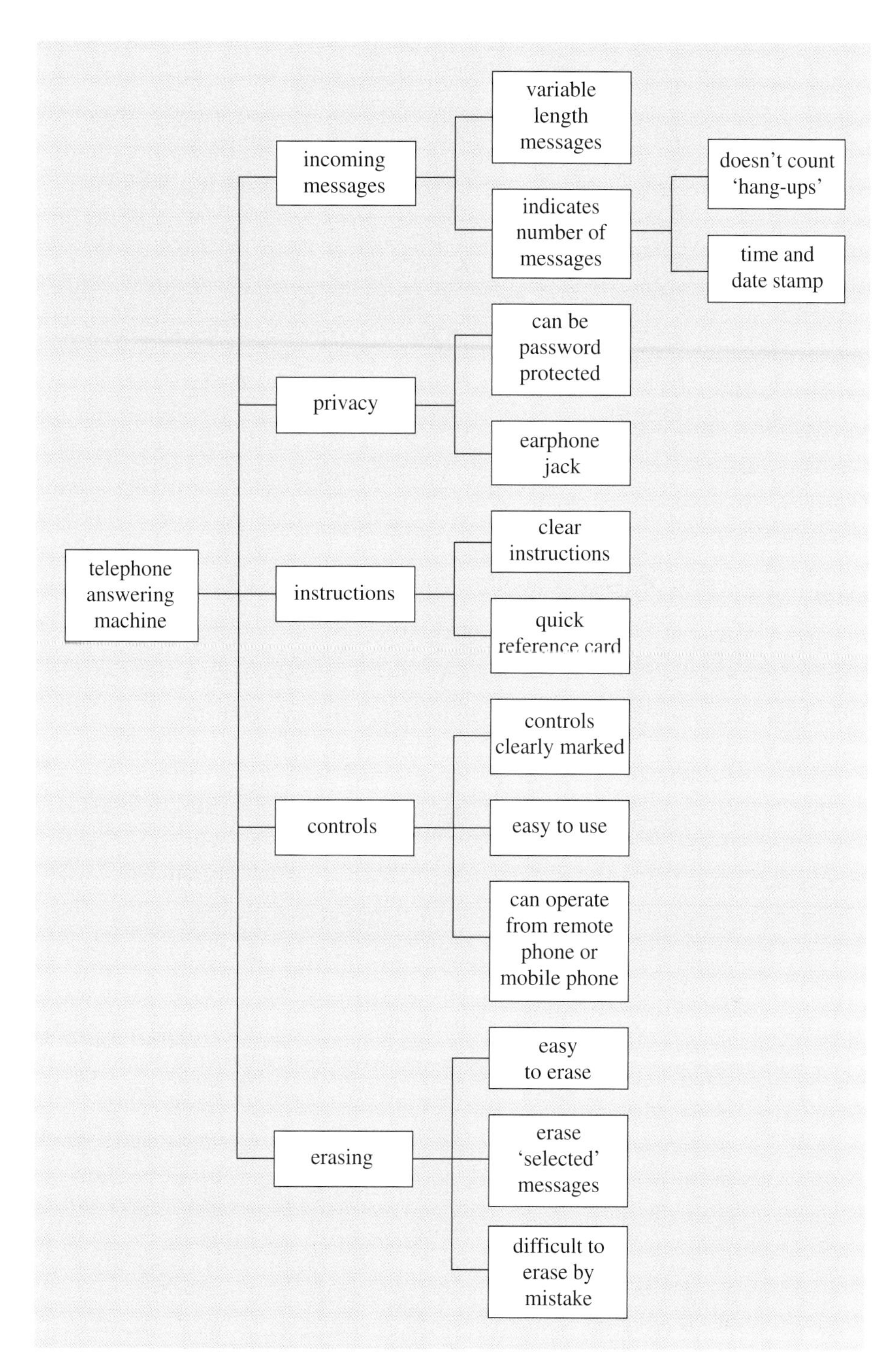

Figure 3.8 A tree diagram for a telephone answering machine

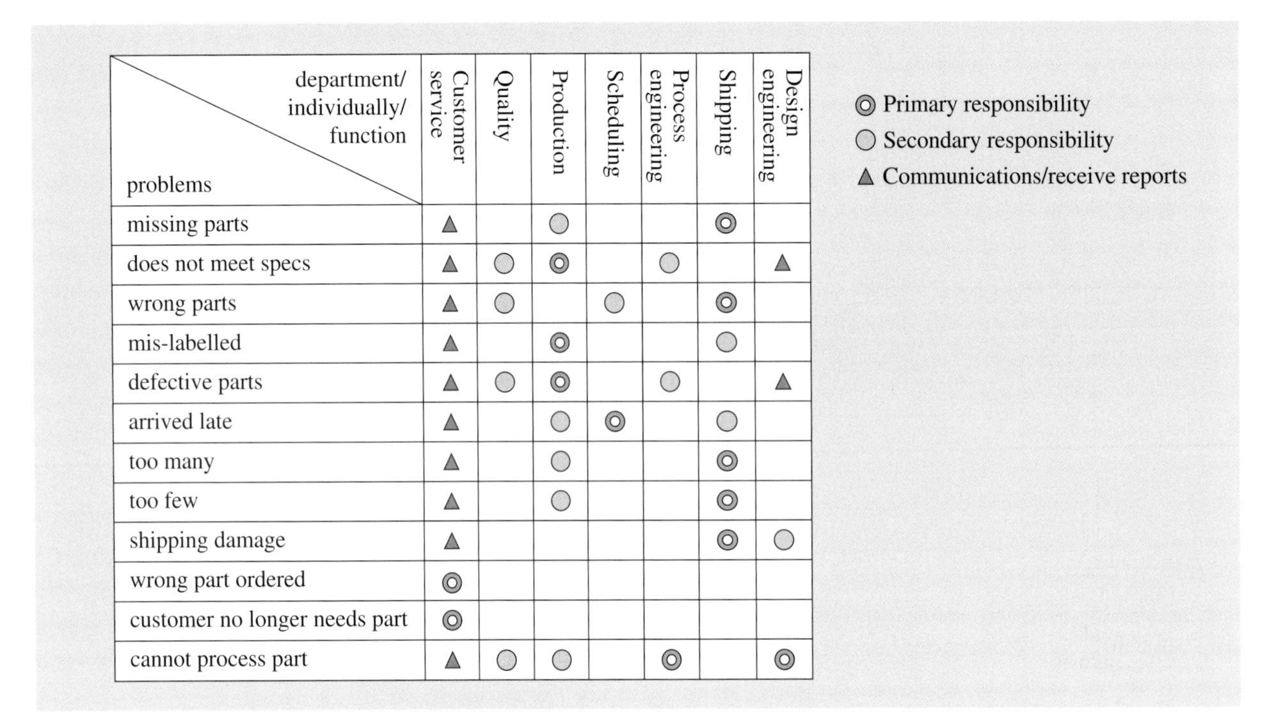

problems \ department/individually/function	Customer service	Quality	Production	Scheduling	Process engineering	Shipping	Design engineering
missing parts	▲		○			◎	
does not meet specs	▲	○	◎		○		▲
wrong parts	▲	○		○		◎	
mis-labelled	▲		◎			○	
defective parts	▲	○	◎		○		▲
arrived late	▲		○	◎		○	
too many	▲		○			◎	
too few	▲		○			◎	
shipping damage	▲					◎	○
wrong part ordered	◎						
customer no longer needs part	◎						
cannot process part	▲	○	○		◎		◎

◎ Primary responsibility
○ Secondary responsibility
▲ Communications/receive reports

Figure 3.9 A matrix diagram looking at delivery problems (Source: Bossert, 1991, p. 87)

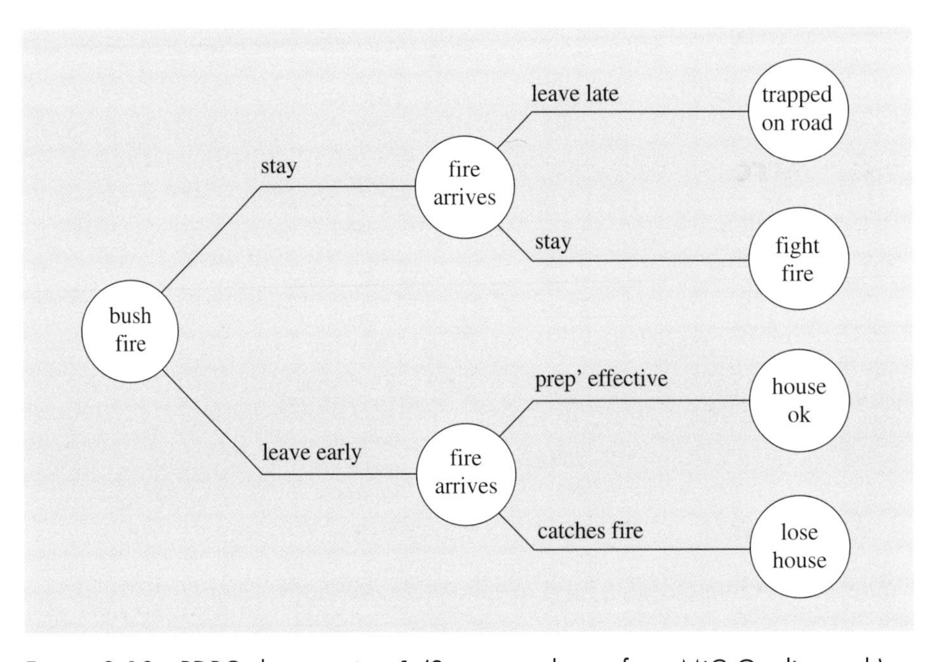

Figure 3.10 PDPC chart version 1 (Source: redrawn from MiC Quality, n.d.)

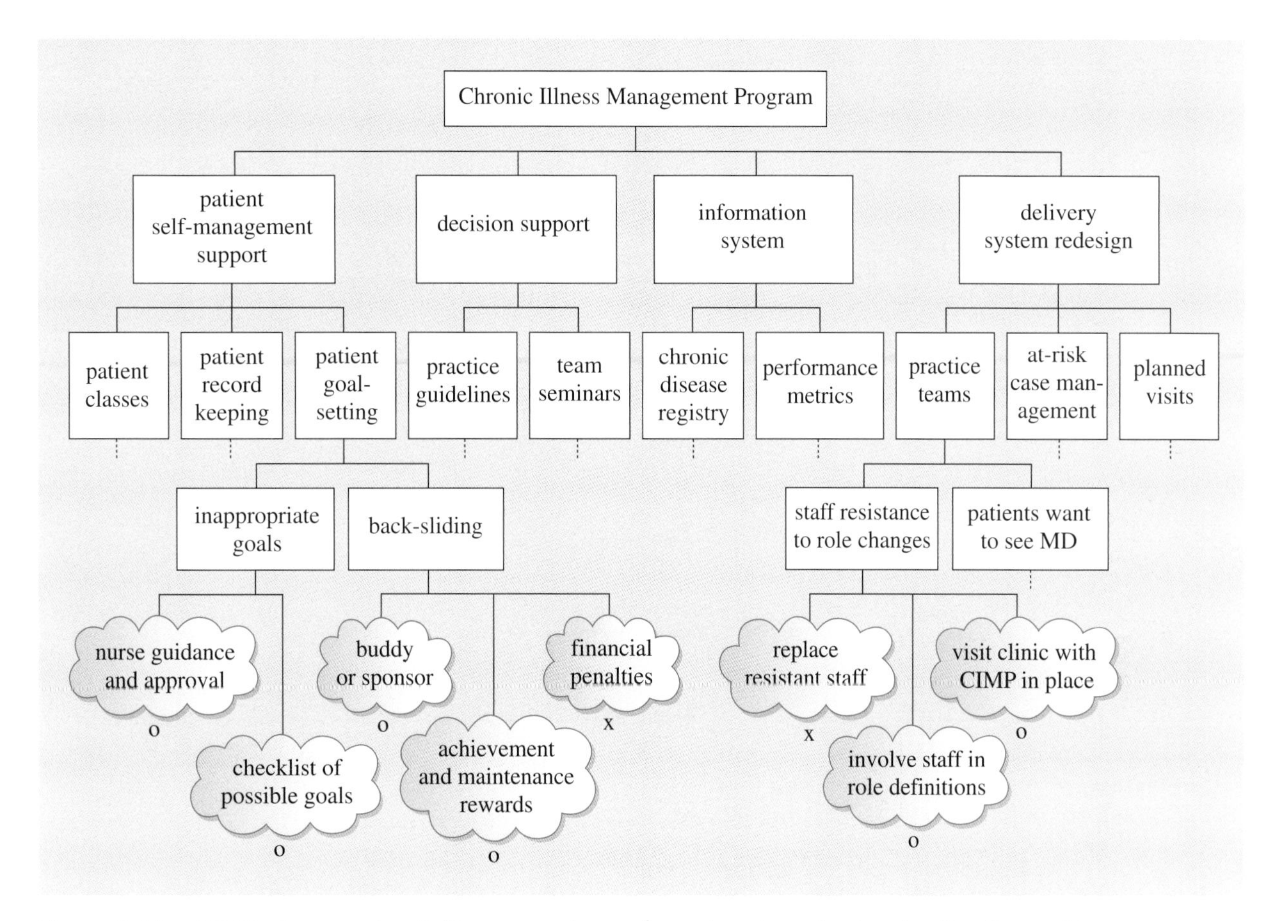

Figure 3.11 PDCP chart version 2 (Source: redrawn from American Society for Quality, n.d.)

3.6 Arrow diagrams

An arrow diagram shows the main structural features of a situation and the important relationships that exist between them. One of the major uses of arrow diagrams is to explore interrelationships with a view to identifying problem areas more precisely and starting to generate possible solutions. Figure 3.12 is an arrow diagram that explores the multiple causes that have led to the collapse in market share of one of a food manufacturer's major lines.

3.7 Matrix diagram analysis

Matrix diagram analysis uses the information about relationships that is revealed by the patterns and by column and row totals on matrix diagrams. A technique that draws on this tool particularly heavily is quality function deployment, which is described in Section 10.

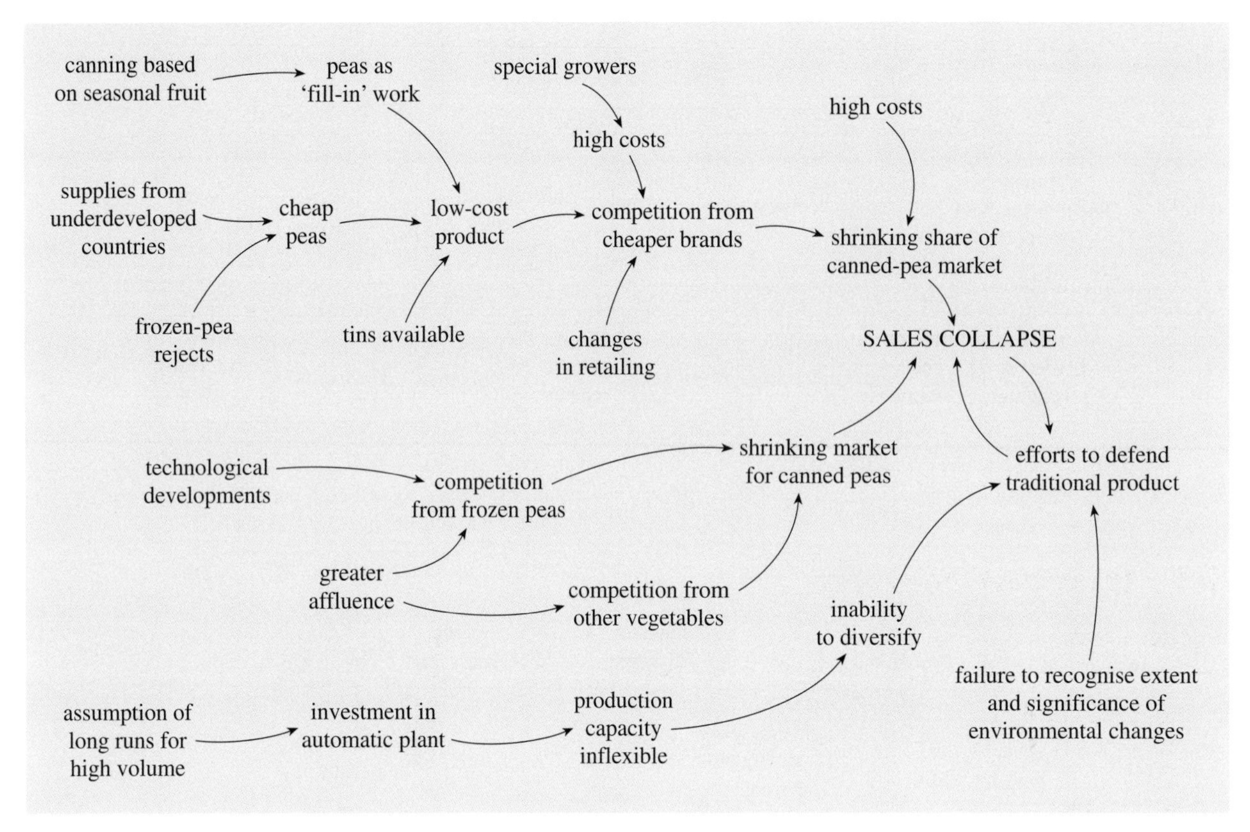

Figure 3.12 Arrow diagram exploring collapse of sales

4 QUESTIONING TECHNIQUES

Questioning techniques are most useful at the earlier stages of problem solving and improvement because they help you to answer the question 'where are we now?'. The techniques that will be covered in this section are:

- is/is not analysis
- five whys
- five Ws and H
- questions about customers
- questions about suppliers
- questions for improving the design of products and processes.

4.1 Is/is not analysis

In general terms is/is not analysis is a way of narrowing options down by asking 'what is it?' and 'what is it not?' It comes in a variety of guises and is often used to determine the scope of an improvement project. A couple of examples are shown in Tables 3.2 and 3.3.

Table 3.2 One form of an is/is not analysis

	Is	**Is not**	**So ...**
What	Product DEF	Product DEF for client A, product GHJ and HIJ	
Where	Team ABC Coventry	Team ABC Liverpool	Check that same processes are used
When	All 1999 results	Results pre-1999	Obtain results for 1999, check that there are sufficient
Who	Only team ABC		
Scope	Will concentrate on team ABC in Coventry and the way they deal with product DEF (excluding that supplied to client A) during the year 1999		

(Source: Marconi, 2000, p. 68)

Table 3.3 Another form of an is/is not analysis

What is area of the problem?	What is not area of the problem?
What are symptoms?	What are not symptoms?
When is the problem observed?	When is the problem not observed?
Where does the problem occur?	Where does the problem not occur?
Who is affected by the problem?	Who is not affected by the problem?

There are two dangers in defining the scope of an improvement project: making it too wide, and making it too narrow. Is/is not analysis cannot prevent either of these but it can ensure that the people working on the project are clear about the decision. It can also serve as a reference point during the project if the scope seems to be extending (known as creep) when it should not, or if the scope needs to be widened.

4.2 Five whys

One of the simplest questioning techniques is known as five whys. It is said to be a favourite of the Toyota executive, Taiichi Ohno. As the extract in Box 3.1 shows, its aim is to drill down to the real problem or to identify the root causes of a problem by continuing to ask the question 'why?'.

BOX 3.1 EVOLUTION OF THE TOYOTA PRODUCTION SYSTEM

Repeating ***why*** *five times*

WHEN CONFRONTED WITH a problem, have you ever stopped and asked *why* five times? It is difficult to do even though it sounds easy. For example, suppose a machine stopped functioning:

1. *Why* did the machine stop?
   There was an overload and the fuse blew.
2. *Why* was there an overload?
   The bearing was not sufficiently lubricated.
3. *Why* was it not lubricated sufficiently?
   The lubrication pump was not pumping sufficiently.
4. *Why* was it not pumping sufficiently?
   The shaft of the pump was worn and rattling.
5. *Why* was the shaft worn out?
   There was no strainer attached and metal scrap got in.

Repeating *why* five times, like this, can help uncover the root problem and correct it. If this procedure were not carried through, one might simply replace the fuse or the pump shaft. In that case, the problem would recur within a few months.

To tell the truth, the Toyota production system has been built on the practice and evolution of this scientific approach. By asking *why* five times and answering it each time, we can get to the real cause of the problem, which is often hidden behind more obvious symptoms.

'Why can one person at Toyota Motor Company operate only one machine, while at the Toyoda textile plant one young woman oversees 40 to 50 automatic looms?'

By starting with this question, we obtained the answer 'The machines at Toyota are not set up to stop when machining is completed.' From this, automation with a human touch developed.

To the question 'Why can't we make this part using just-in-time?' came the answer 'The earlier process makes them so quickly we don't know how many are made per minute.' From this, the idea of production leveling developed.

The first answer to the question 'Why are we making too many parts?' was 'Because there is no way to hold down or prevent overproduction.' This led to the idea of visual control which then led to the idea of kanban.

... the Toyota production system is based fundamentally on the absolute elimination of waste. Why is waste generated in the first place? With this question, we are actually asking the meaning of profit, which is the condition for a business's continued existence. At the same time, we are asking why people work.

In a production plant operation, data are highly regarded – but I consider facts to be even more important. When a problem arises, if our search for the cause is not thorough, the actions taken can be out of focus. This is why we repeatedly ask *why*. This is the scientific basis of the Toyota system.

(Ohno, 1988, pp. 17–18, originally published in Japan, 1978)

4.3 Five Ws and H

Instead of just asking 'why?' at different levels of a problem five Ws and H also asks 'who?', 'what?', 'where?', 'when?' and 'how?'. The technique thus provides a systematic framework for fact finding. An alternative name for it is the Kipling checklist. It gets this name from one of the poems in Kipling's *Just So Stories*:

> I keep six honest serving-men
> (They taught me all I knew);
> Their names are What and Why and When
> And How and Where and Who.
>
> (Kipling, 1902)

Federico takes it one stage further by establishing a subset of questions for four of the Ws and the H in order to move forward from fact finding to the generation of potential improvements:

What?	What happens now? Why do it? Can we do something else?
Who?	Who does it? Why them? Can someone else do it?
Where?	Where is it done? Why there? Can we do it elsewhere?
When?	When is it done? Why then? Can we do it some other time?
How?	How is it done? Why this way? Can we do it some other way?

(Federico, 2005, p. 162)

4.4 More specific sets of questions

In looking for improvements, more specific questions can provide a good starting point for finding out where you are. I have included three examples in this section. They begin with the concept that is often central to improvement: satisfying the customer.

Questions about customers (internal and external)

- Who are our immediate customers?
- What are their true requirements?

- How do we or can we find out what the requirements are?
- Do we have the necessary capability to meet the requirements? If not, what must change to improve the capability?
- Do we continually meet the requirements? If not, what prevents this from happening when the capability exists?
- How do we monitor changes in the requirements?

Questions about suppliers

- Who are our immediate suppliers?
- What are our true requirements?
- How do we communicate our requirements?
- Do our suppliers have the capability to measure and meet the requirements?
- How do we inform them of changes in the requirements?

Questions for improving the design of products and processes

These are shown in Table 3.4.

Table 3.4 Questions for improving the design of products and processes

	Product	**Process**
Eliminate	Can any of the components be eliminated?	Can any of the activities be eliminated?
Integrate	Can one component be integrated with another component?	Can one activity be integrated with another activity?
Combine	Can the given components be combined in a better way?	Can a better sequence of activities be followed?
Simplify	Can components be simplified?	Can activities be simplified?
Standardize	Can components be standardized into one?	Can activities be standardized into one?
Substitute	Can any component be replaced?	Can any activity be replaced?
Revise	Can any component be revised?	Can any activity be revised?

(Source: Booker et al., 2001, p. 275 from Huang, 1996)

The purpose of questioning techniques is to draw out all relevant information. It is important to record it all in an inclusive, open and non-judgemental way in the first instance and to accept that the answers you get may vary depending on whom you ask! One of the weaknesses of such techniques is that they tend to produce a lot of information that is not very well integrated and is not good at revealing interconnections between the answers to separate questions. In the next section I shall look at a set of techniques that emphasises the whole and the relationships between the parts.

5 SYSTEMS DIAGRAMS

The techniques that will be covered in this section are:

- input-output diagrams
- systems maps
- influence diagrams
- rich pictures.

The notion of 'system' lies at the heart of a very useful set of concepts and techniques for operating on problems or developing opportunities. I shall concentrate here on those that can be used to investigate and describe a current situation. In these techniques, system can be defined as an 'organised whole' or a set of components that are interconnected and working together to achieve a purpose. Because system can be used as an abstract notion and applied to any situation, not just one that would popularly be labelled as a system, it is a very useful vehicle for gaining understanding of any situation.

The need for systems ideas was felt initially in the process industries where the large-scale integration of plant led to severe problems of design, management and maintenance. For instance, a production-planning problem of immense complexity was the effective operation of a group of oil refineries, using several grades of crude oil from different parts of the world to make products ranging from heavy fuel oil to aviation spirit with varying patterns of seasonal demand. Similarly, the integration of the UK electricity supply industry after nationalisation in 1948 created a large physical network which needed a sophisticated approach to system design and to operating and control strategies. Problems like these can be found in both service and manufacturing organisations, particularly now that the complexity of services and products, the intensity of competition, and the diversity of markets have all increased. For example, a major food retailer faces the challenge of coping with thousands of different lines and all the problems for purchasing, scheduling and delivery that they entail. In addition, the use of information technology to process information relating to operations has increased, rather than reduced, the need to take an overall view of the operations concerned, and has itself added to the complexity.

Systems concepts and techniques, such as diagramming and other forms of modelling, make it easier to understand complexity. By recognising the importance of outside influences, and by breaking down what is perceived as the overall system into interacting interdependent subsystems, a better understanding of a situation and its problems can be achieved. For example, a problem expressed as a shortage of space in a warehouse might lead to a conventional solution that involves extending the warehouse or increasing its capacity by reorganising it internally. A systems approach, on the other hand, with its emphasis on holism, could regard the warehouse as a buffer system between production or receipt of incoming goods and dispatch and, by examining the planning and scheduling system and the decisions related to the dispatch of goods, might suggest that reducing the stock level would provide a better solution.

When considering aspects of a situation as a system it is useful to draw a distinction between structure and process. The structure of a system comprises those features which do not change, or change only slowly over time. Processes occur in the system and are either concerned with the transformation of inputs into outputs or are associated with the maintenance of the system.

In looking at the structure of a system it is necessary to identify the system's components. The smallest part of a system that it is helpful to consider is called an element. If groups of elements within a system seem to be linked together in some organised way they are said to form subsystems. For example, if you were to have a supplier as part of a system you could regard the supplier as a subsystem made up of a production facility, workforce, management, and so on. Whether you chose to regard it as a single component labelled supplier or as a subsystem made up of a number of components would depend on the purpose of your study.

The components within a system are said to lie within a system boundary. The boundary can be regarded as the imaginary line delineating that which is considered to be part of the system from that which is not part of the system. The environment of the system lies outside the boundary. This is not everything in the universe that is not part of the system, or something to do with the weather, but simply those things that influence the behaviour of the system or are influenced by it. The components of the environment can also exert a degree of control over the system but the environment cannot be controlled by the system. Take the example of a system that has mobile phones as one of its components in order to allow communication to take place. The masts transmitting signals to the phone would probably not be part of the system but they would be part of the system's environment. If they were out of action the system would not be able to communicate.

Another important systems concept is hierarchy. Suppose you are studying a particular system. If it would be helpful, you could take different views: the system you are currently considering could be regarded as a subsystem of a wider system; or a subsystem of the current system could be regarded as a system. Systems can thus be seen as nested together and interconnected in a way that can be likened to that of a software system with a suite of programmes within it.

The hierarchical nature of systems within systems and the iterative way in which they interconnect is an important feature of systems thinking. It leads to systems producing behaviours that are either more or less than the sum total of their individual components. As Vickers puts it:

> It is less, because organization constrains. Elements in a system are not free to do all the things which, unorganized, they might do. It is more because, when organized they are enabled to do together what none of them could do alone, or, if unorganized, even together.
>
> (Vickers, 1981, p. 21)

The idea of the whole being more, or less, than the sum of the parts is given the name 'emergence'. Sometimes emergence is defined by reference to different hierarchical levels of organisation where each particular level is 'characterised by emergent properties which do not exist at the lower [less complex] level' (Checkland, 1981, p. 78). Other definitions draw more general attention to the notion of interaction as a cause of emergence. For example, Casti states that emergence is:

> an overall system behaviour that comes out of the interaction of many participants – behaviour that cannot be predicted or even envisioned from a knowledge of what each component of the system does in isolation.
>
> (Casti, 1997, p. 82)

Hebel provides the following definition:

> An emergent property of any system is the result of a collective interaction of components, although it is often an unintended consequence. It can be singular or multiple and have positive and negative effects.
>
> (Hebel, 1999, p. 257)

Emergent properties can be linked to system behaviour. For example, stability may be an emergent property of a system incorporating effective feedback control. Emergent properties also have an important effect, positive or negative, on system outputs. For example, a system with commercial objectives, made up of people, materials, skills and equipment, ought to have as an emergent property 'ability to respond to market demands'. If it has not, then the notion of emergent property may help to develop a potentially important insight into a problem.

In investigating complex situations, drawing systems diagrams is a valuable way of clarifying thinking and of conveying information about the situation to other people. Systems diagrams can capture and represent interconnectedness in a way that is usually impossible in linear text. Thus diagramming is an essential component of nearly all systems work.

5.1 Input-output diagrams

An input-output diagram is a simple form of flow-block diagram with the system represented by a single box, and inputs and outputs shown as labelled arrows. An example is shown in Figure 3.13. Although this diagrammatic form is very simple it is a very effective way of showing what the system would or should do, that is, take inputs and transform them into outputs. Answers to the questions 'Who are my suppliers?' and 'Who are my customers?' (see Section 4.4) can be used to help to construct this diagram, but inputs and outputs are usually broader than inputs from suppliers and the products and services delivered to customers. For example, labour is usually an important input but only in special cases such as the use of an agency to provide temporary staff would you identify labour suppliers.

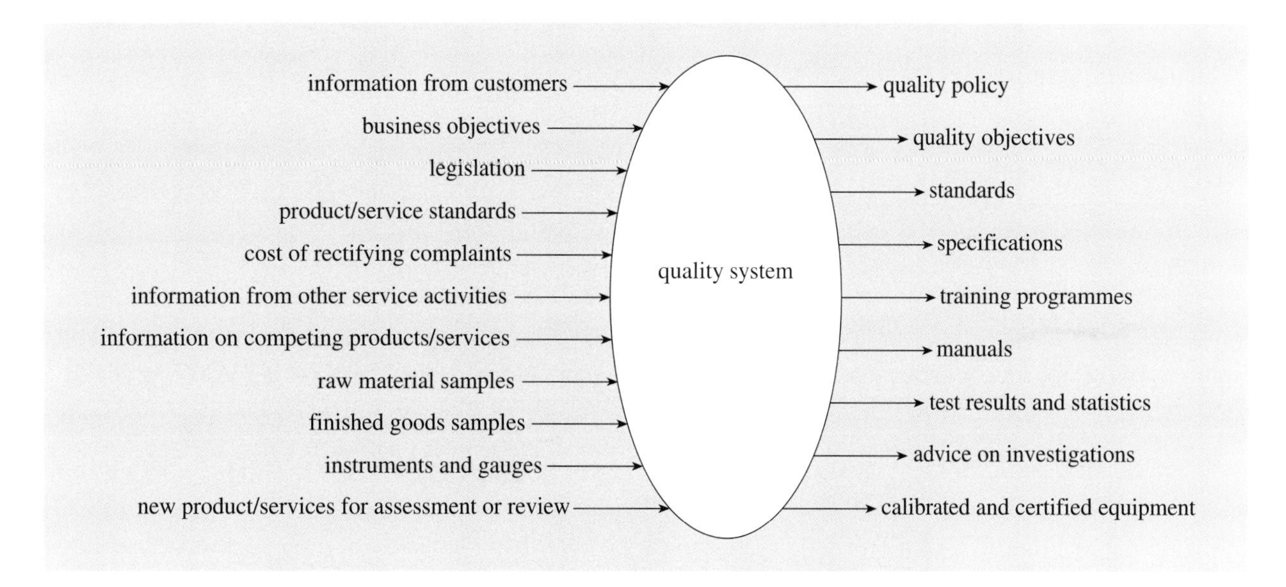

Figure 3.13 Example of an input-output diagram

ACTIVITY 3.3

Suggest three inputs and three outputs of a quality improvement system.

If you want to focus on suppliers and customers instead of the broader categories of inputs and outputs, then a useful way of expressing the system is as a total supply network. An example is shown in Figure 3.14. ●

5.2 Systems maps

A systems map is essentially a snapshot showing the components of a system and its environment at a point in time. An example is shown in Figure 3.15. Primarily it shows structure, but the positioning of components provides some information about the relative strengths of relationships.

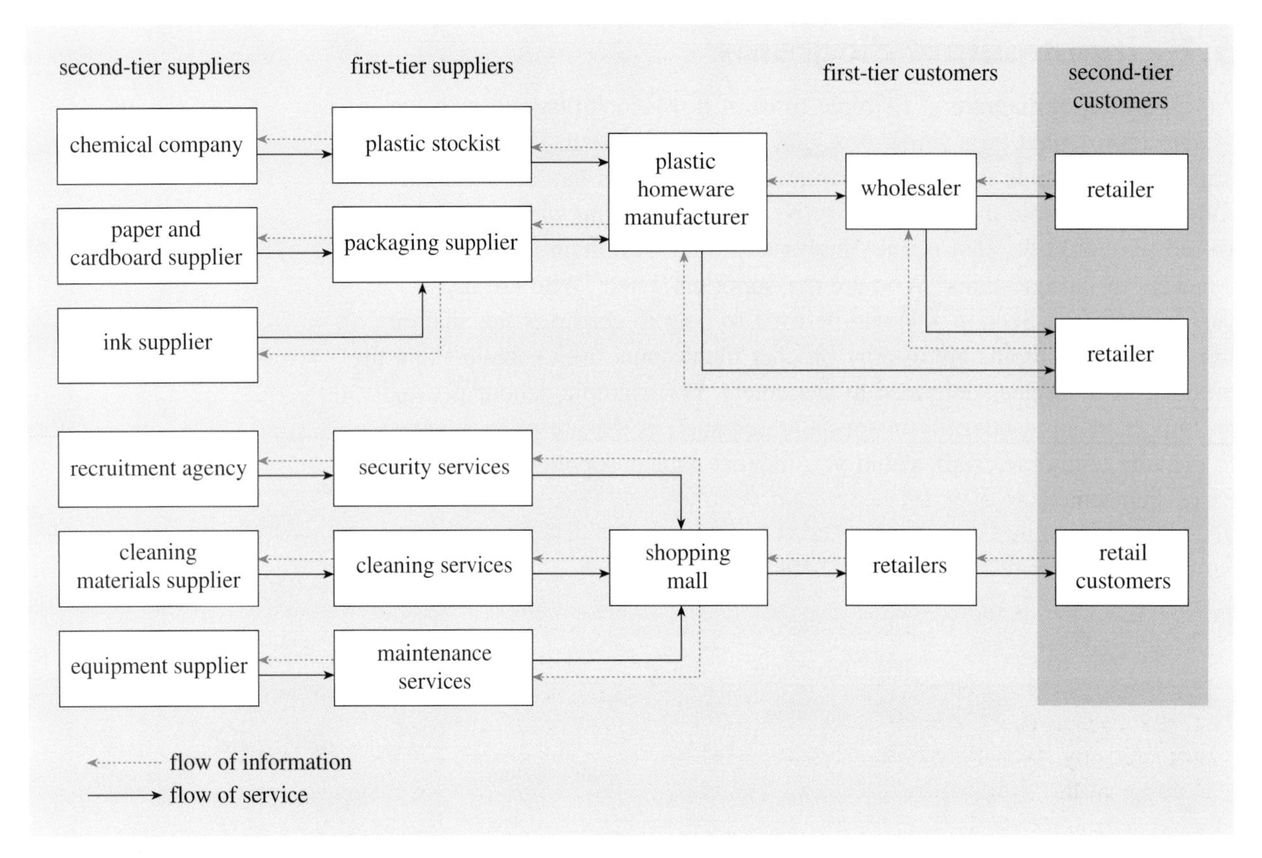

Figure 3.14 Total supply network for a shopping mall (Source: redrawn from Slack et al., 2007, p. 149)

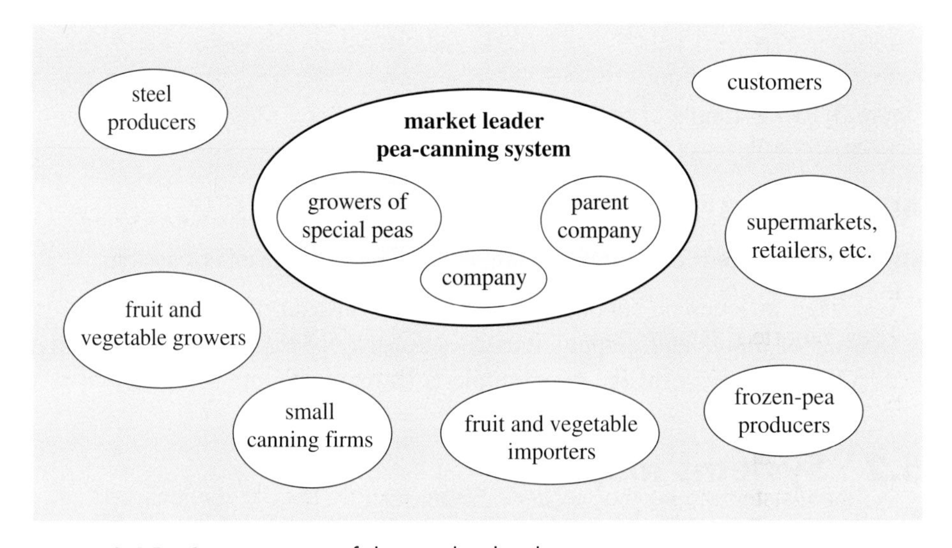

Figure 3.15 Systems map of the market leader pea-canning system

Where systems share common components, overlaps between subsystems may occasionally need to be included, but their use should be kept to a minimum. The following are helpful conventions in drawing systems maps:

1 Lines represent boundaries around systems, wider systems, subsystems or components.
2 Each 'blob' formed by a boundary should be labelled. Beware of labels that relate to properties, variables, levels, value judgements, and so on, rather than to the names of systems or components.
3 Overlaps occur only when a component is a member of more than one system, and should be kept to a minimum.

ACTIVITY 3.4

Read the scenario in Box 3.2 and then draw a systems map to help you understand the situation described there. Give a name to the system and show the system within its environment. ●

BOX 3.2 CHILDREN AT RISK

The Director of Social Services in a County Council became concerned at the recent cases where children at risk had not been adequately protected by social workers. He requested data on the numbers of children in the county:

- who had been taken into care
- who were being visited by social workers regularly
- who might be in danger according to reports from members of the public.

When he received all the figures under each category, he compared them with the files and reports and found that the numbers simply didn't add up. That is, in some categories the files and reports suggested that there were more children involved than were referred to in the data, and in some cases there seemed to be fewer.

Deeply concerned at this state of affairs, he called a meeting of the six area heads of social services. The outcome was an agreement that the information systems were inadequate. A review was put in hand by the management services division.

They reported that the basic procedures for collecting information were adequate, but that the social workers who provided the raw data were both:

1 sloppy about providing the data on the due dates;
2 inconsistent about the categories they used to classify the data.

To try and solve the problem, they produced a set of draft rules on data collection for social workers.

The Director called a meeting with a representative group of social workers to discuss the draft rules. They reacted fiercely. They said that the social workers were under great stress owing to increased workloads, reduced resources, the great responsibility they had for the children and the general lack of appreciation for their work. It was hard enough to do the job as it was, but if they had to spend even more time reading directives and filling in forms, the situation would become intolerable. In any case, their clients weren't just numbers; they were individual people with individual problems. Data like this would probably be used to prove they got more resources than they should have, according to some bureaucratic guideline.

5.3 Influence diagrams

An influence diagram is also a snapshot based on structure, but its prime purpose is to explore relationships. It is often the case that closer examination of interactions leads to redefinition of the system or regrouping of its components.

The elements of an influence diagram are named components and arrows. The arrows may be labelled to distinguish between different types of influence such as influence via finance, information, or statutory regulation, and variations in line thickness may be used to indicate different strengths of influence. As in systems maps, space and relative distance reveal information about the nature of relationships: for example, a component that is shown as having a strong influence could also, from its position, be seen as isolated and remote.

Figure 3.16 is an influence diagram developed from the systems map in Figure 3.15. A one-way arrow such as that joining the supermarkets,

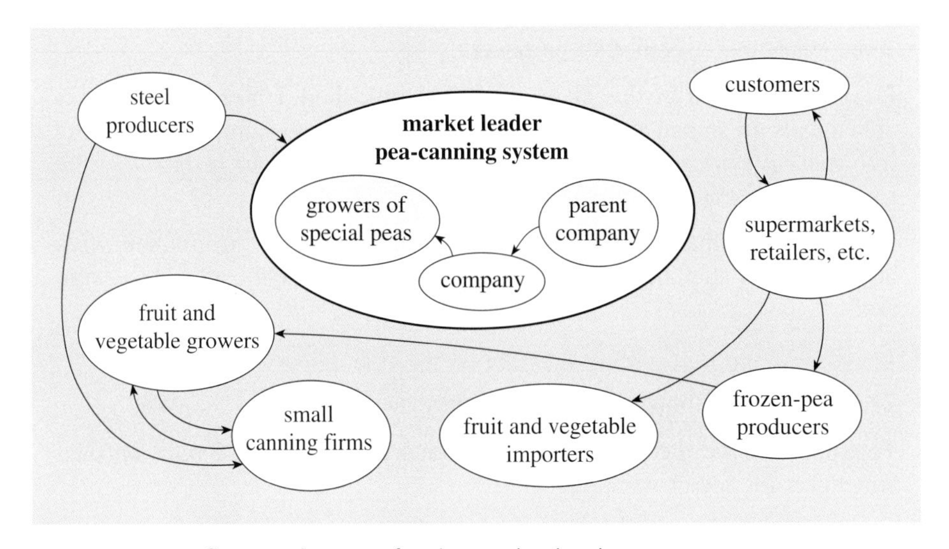

Figure 3.16 Influence diagram for the market leader pea-canning system

retailers, etc. to the frozen-pea producers shows that the former can or do influence the latter, but the latter has no significant influence over the former. Double-headed arrows can be used to denote two-way influences but it is preferable to use two separate opposing arrows.

ACTIVITY 3.5

Using the systems map given at the end of this block as the answer to Activity 3.4 (Figure 3.61), draw an influence diagram to help you understand the relationships in the situation. ●

5.4 Rich pictures

Although a rich picture is not a representation of a system I have included it in this section because of its association with Checkland's soft systems methodology (SSM) (Checkland, 1972). (SSM is dealt with in Block 4.) Some people interpret the term 'rich picture' literally, specifying that it must be a physical picture, while others consider it to refer to an abstract understanding of a problem situation (see Lewis, 1992). A diagrammatic rich picture seeks to get on to one sheet of paper all the salient features of a situation in a way that is insightful and can be easily assimilated. It is common to use cartoon-like encapsulations of key ideas or pieces of information, as in the example in Figure 3.17 which shows a rich picture depicting the situation at a large engineering firm (SEM) with significant supply chain management problems.

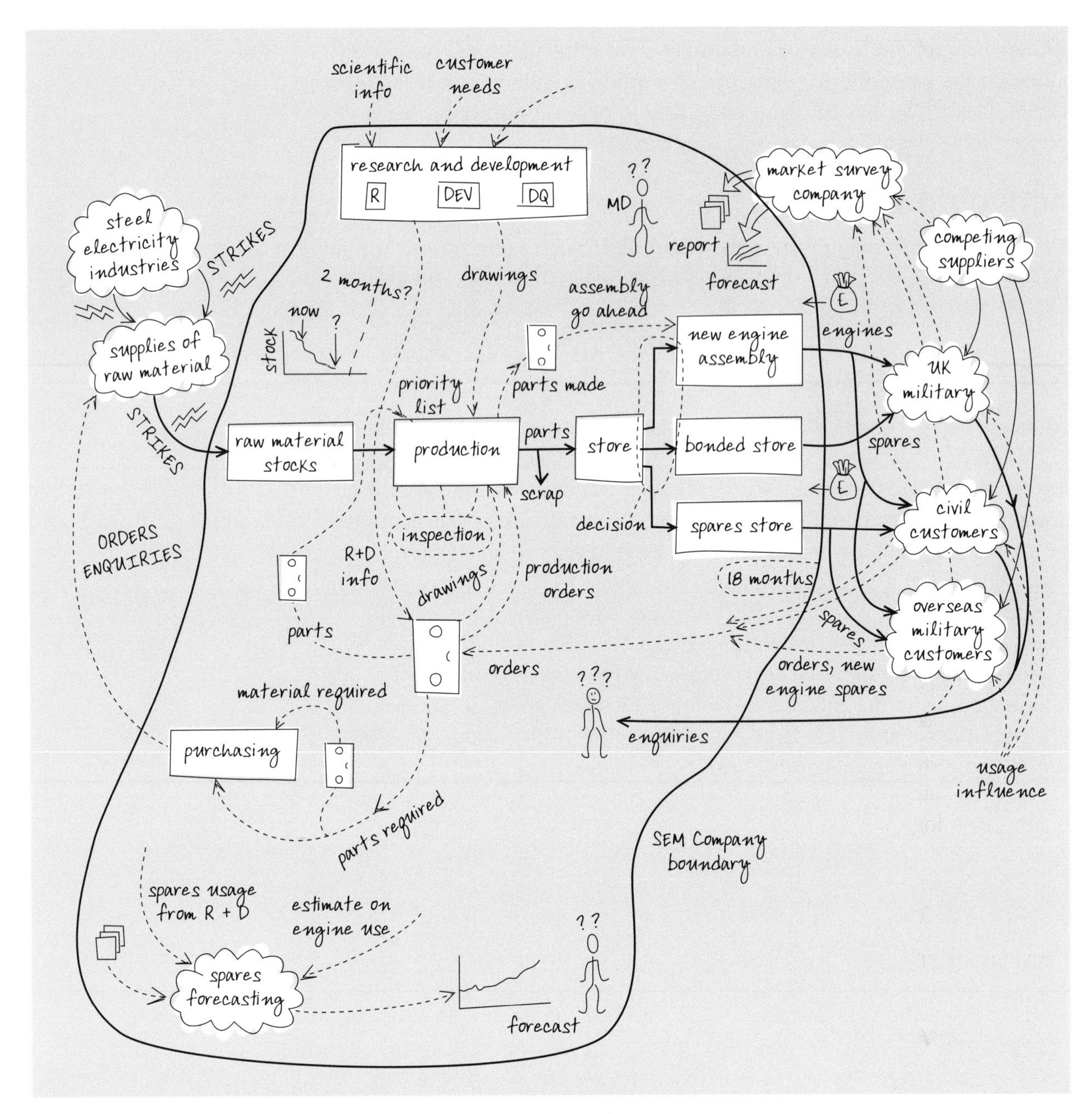

Figure 3.17 Rich picture of the situation at an engineering firm, SEM

6 OTHER DIAGRAMMING TECHNIQUES

The techniques that will be covered in this section are:

- activity sequence diagrams
- flow-block diagrams
- flow-process diagrams
- spaghetti charts
- SIPOC charts
- multiple-cause diagrams
- force field analysis
- cognitive mapping

6.1 Activity sequence diagrams

Activity sequence diagrams can be used either to look at the activities being carried out in existing situations or to generate solutions and improvements by modelling the activities that need to be carried out in order to achieve desired outcomes. Once you know what activities have to be undertaken you can look at the structures and processes that are necessary to allow these activities to take place.

Figure 3.18 shows an activity sequence diagram for a mail order processing system. The arrows indicate either time elapsing between a sequence of actions or a logical necessity for one activity to end before the next one can begin.

ACTIVITY 3.6

Draw an activity sequence diagram for a car rescue service. ●

6.2 Flow-block diagrams

Flow-block diagrams are useful for describing how materials, information or energy flow between the components of a system. The principal conventions in drawing flow-block diagrams are as follows:

1. An arrow should represent a flow of material, energy or information from one component to another.
2. Labels, colours or a key should be used to identify the flows.
3. Boxes should represent system components or processors and should be labelled.
4. The output(s) from a box should be a transformation of the input(s) to the box.

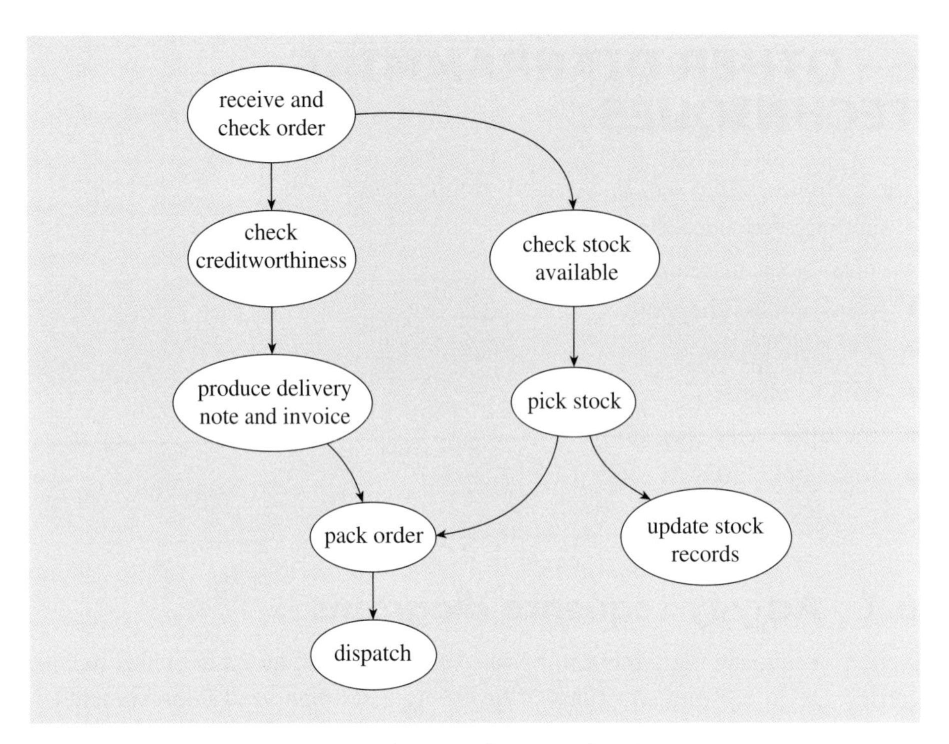

Figure 3.18 Activity sequence diagram for a mail order processing system

6.3 Flow-process diagrams

Flow-process diagrams are closely related to flow-block diagrams, but here the blocks identify the processes rather than the components or processors. For a process that is producing a service rather than a product, a flow diagram may depict the progress of something less tangible than material such as information, or show the customer moving through the process. Figure 3.19 shows a flow-process diagram for a mail order processing system.

A useful way of distinguishing flow-block diagrams from flow-process diagrams is to remember that in flow-block diagrams the boxes represent stages and are labelled with nouns, whereas flow-process diagrams are constructed in terms of activities and the boxes are labelled with verbs. Thus, the box in Figure 3.19 labelled 'pack order' would be labelled 'packing' in a flow-block diagram. The two conventions should not be used together in one diagram.

ACTIVITY 3.7

Draw a flow-process diagram for a car rescue service. ●

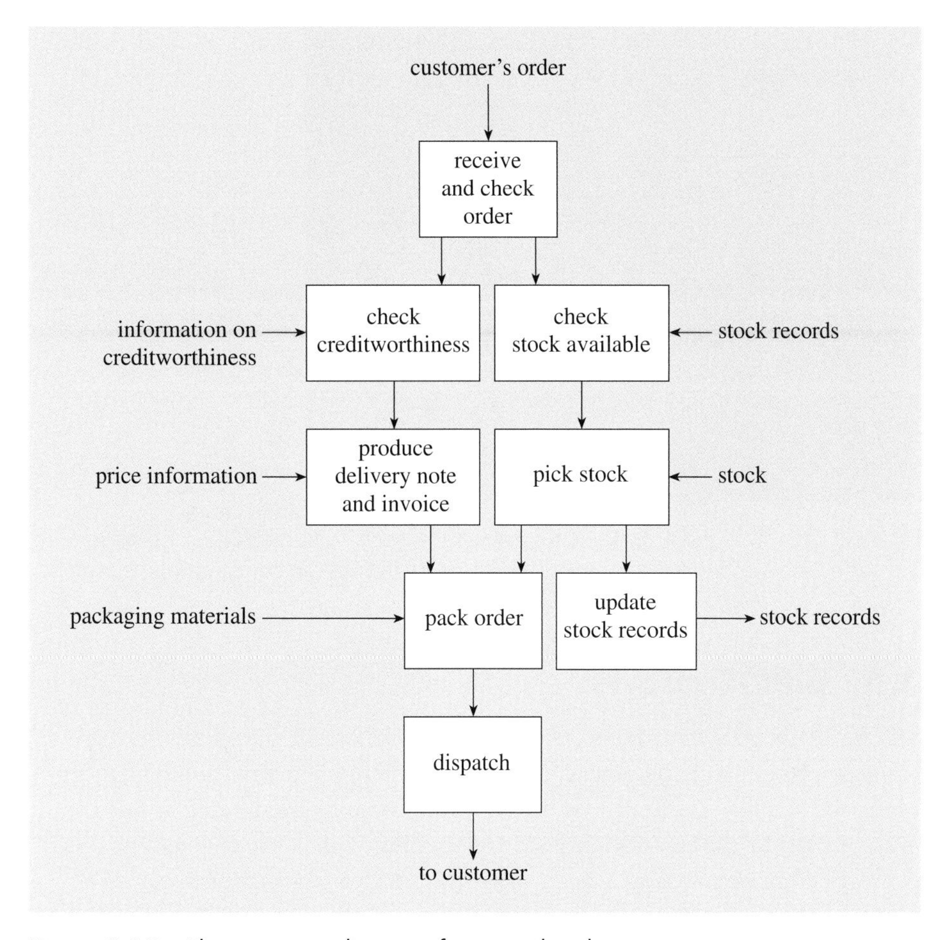

Figure 3.19 Flow-process diagram for a mail order processing system

6.4 Spaghetti charts

Spaghetti charts depict the path followed by people, materials or information in the course of working through a process. An example is shown in Figure 3.20. Features on the completed chart such as crisscrossing lines and repeated returns to the same point can provide evidence of excessive amounts of movement and/or poorly designed layouts.

The information needed to draw a spaghetti chart can be obtained by physically following people, materials or information but a more high-tech method is to use time-elapsed video or photography or GPS (global positioning system).

In order to look for ways of improving process design and the design of physical layouts it is helpful to use spaghetti charts alongside flow-block and/or flow-process diagrams and throughput time analysis and bottleneck analysis (see Section 10.7).

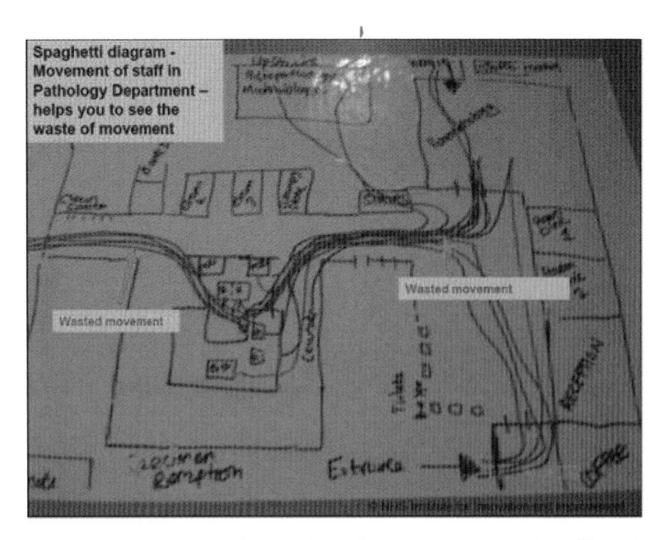

Figure 3.20 Spaghetti chart for movement of staff in the Pathology Department of Hereford Hospital (Source: NHS Institute for Innovation and Improvement, 2006)

6.5 SIPOC charts

SIPOC charts take their name from the initial letters of suppliers, inputs, process, outputs and customers, and thus combine elements from a number of the techniques you have met already. They can be used early in an investigation to capture a high-level representation of a process and its context before conducting a more detailed examination. An example of a SIPOC chart is shown in Figure 3.21.

6.6 Multiple-cause diagrams

You have already met one technique for exploring causes: the cause-and-effect diagram. A more sophisticated technique that can be used instead of a cause-and-effect diagram is the multiple-cause diagram. This is a particular form of arrow diagram that explores why a given event happened or why a certain class of events tends to occur. Its elements are phrases and arrows. Each phrase may be the name of a 'thing', with or without its relevant associated variable(s), or an event, and each arrow may represent a cause or may mean 'contributes to', 'is followed by', 'enables', or something similar. Unless annotation indicates otherwise, no distinction is drawn between necessary and sufficient causes. Construction normally begins at a single factor or event which is then traced backwards.

Figure 3.22 is multiple-cause diagram relating product quality, market and financial factors. It contains loops that feed back on themselves. By following a loop round, it is possible to see whether it is a positive feedback loop or a negative feedback loop. Positive feedback is usually found in stable situations and occurs when a change in a variable produces one or more associated changes that tend to reinforce the original change. In contrast,

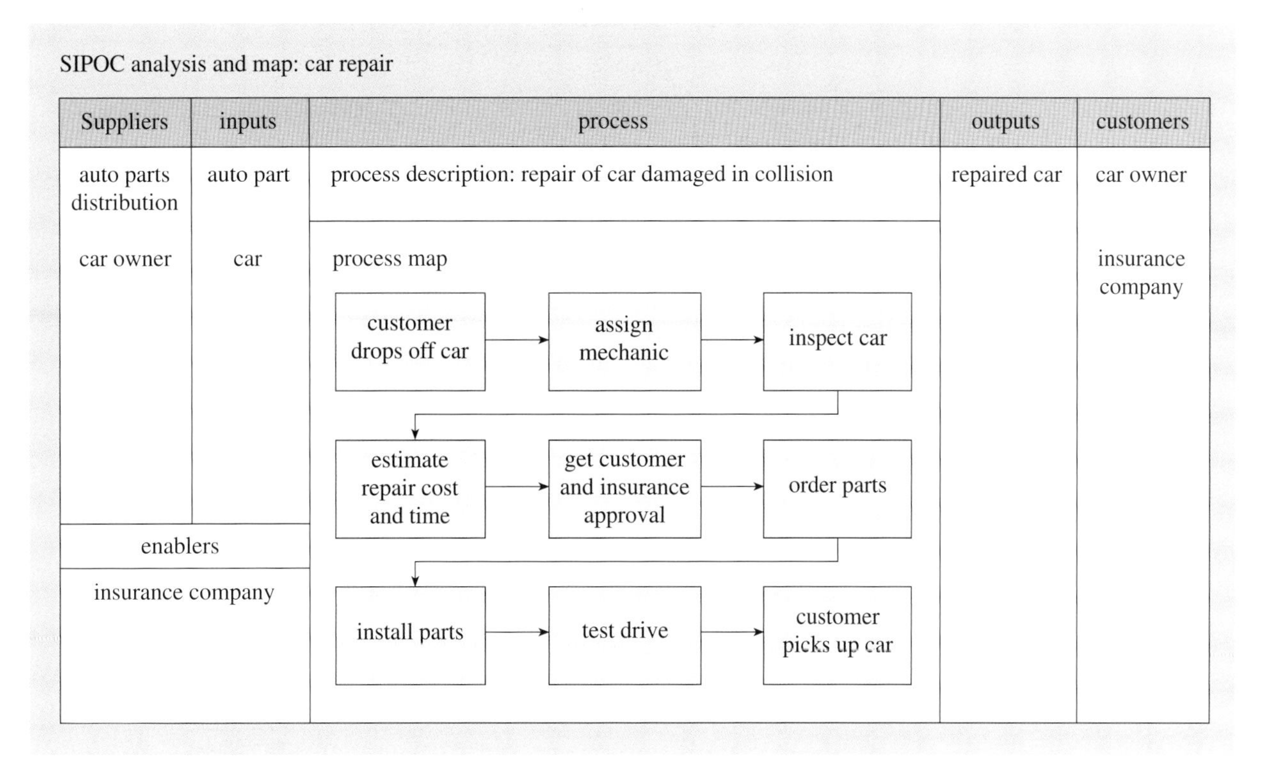

Figure 3.21 SIPOC chart for a car repair process (Source: US Army Enterprise Solutions Competency Center, n.d.)

negative feedback is usually found in unstable situations and occurs when the associated changes feed back to oppose the original change. For example, consider the loop in Figure 3.22 that includes quality of design. Assuming that the quality of design improves, the following sequence of changes might be expected: quality of finished product improves, number of complaints reduces, rectification costs reduce, overall costs reduce, profit increases, resources available for design increase and therefore, it is hoped, the quality of design improves. Thus, the improvement in the quality of design leads to changes that produce further improvement. This positive feedback tends to produce instability, but as growth is one form of instability (as is decline), the existence of some positive feedback loops is to be expected and, as in this case, welcomed.

ACTIVITY 3.8 ....

Use the information in Box 3.2 (Activity 3.4 in Section 5.2) to draw a multiple-cause diagram that examines the data recording problems. ●

6.7 Force field analysis

In almost all situations there are both positive and negative forces at work. The positive forces support the achievement of goals and objectives, high

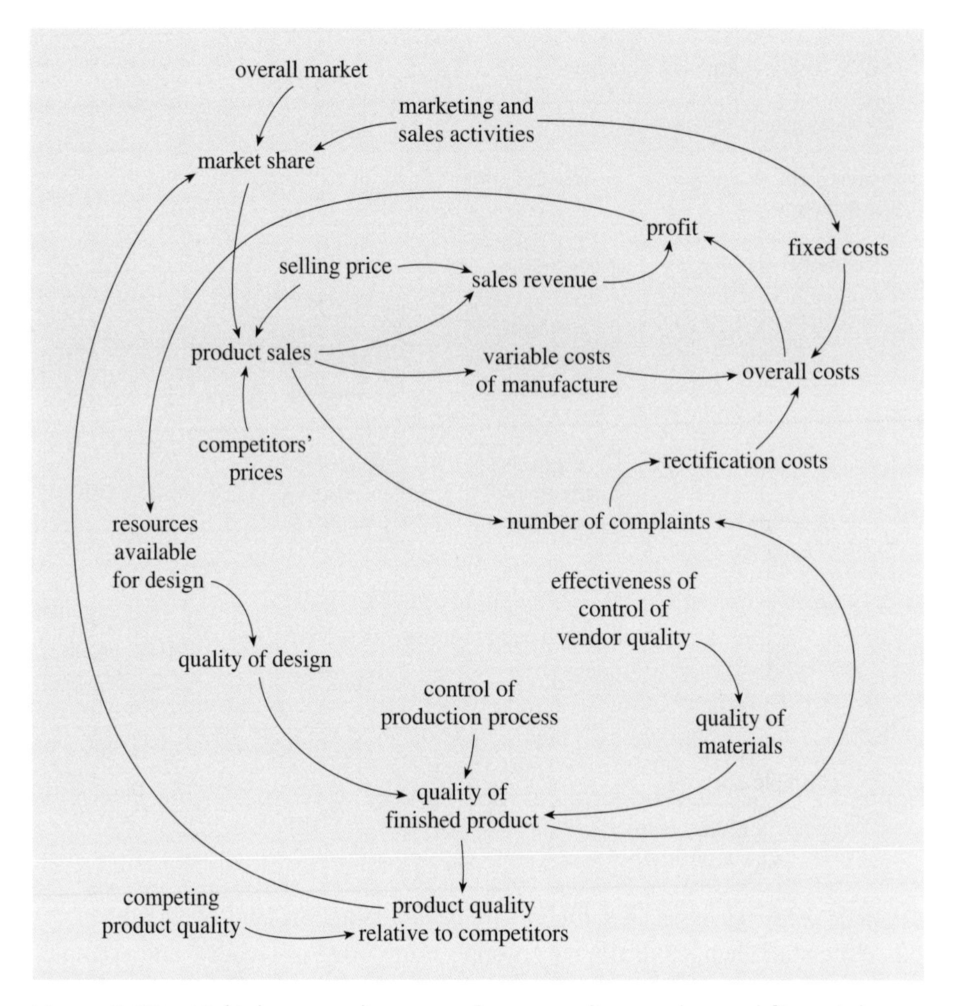

Figure 3.22 Multiple-cause diagram relating quality, market and financial factors

levels of performance and other things that are seen as desirable such as willingness to change and look for improvements; and negative forces push against these things. Some of these forces are exerted from within the situation and others originate in the environment. Force field analysis is a way of encouraging people to identify the forces at work and classify them as driving (positive) forces and restraining (negative) forces. By identifying restraining forces that can be eliminated or weakened and driving forces that can be strengthened, the technique can show where to focus problem-solving and improvement efforts. Two other courses of action that can be very powerful are: changing a restraining force into a driving force – for example, convince a group of people that change is in their interest and foster their commitment to that change; and introducing completely new driving forces.

Force field analysis can be conducted at any level from strategic to individual process, so the first step is to specify the situation that is being examined and the purpose of the analysis. The driving and restraining forces

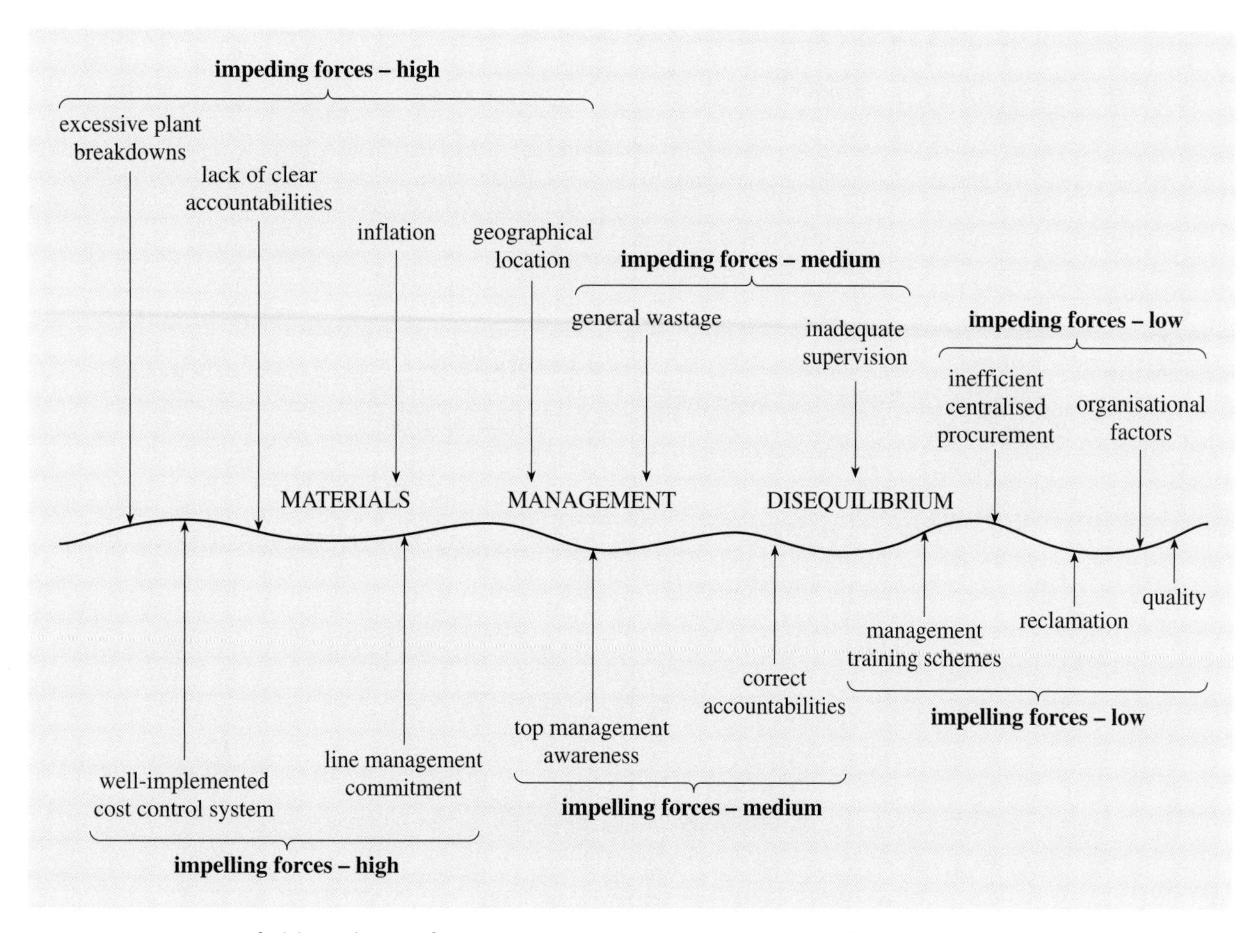

Figure 3.23 Force field analysis – format 1 (Source: Ajimal, 1985. p. 60)

that are currently at work are then identified and the relative strength of the forces is assessed. It is usual to express the findings of the analysis diagrammatically. Figures 3.23 and 3.24 show the two formats that are most commonly used. The second is known as a T-chart.

6.8 Cognitive mapping

The text about cognitive mapping is contained in an offprint.

Now read Offprint 2.

ACTIVITY 3.9

Figures 3.25 and 3.26 show cognitive maps that represent the aggregated views of some multinational biotechnology firms towards the development of genetically modified (GM) crops in relation to their other agrochemical products, for example pesticides. (The figures are taken from a study that looked at how multinational biotechnology companies reacted to European public policies and attitudes to GM.) Figure 3.25 shows how the development of GM crops is perceived as contributing to sustainability

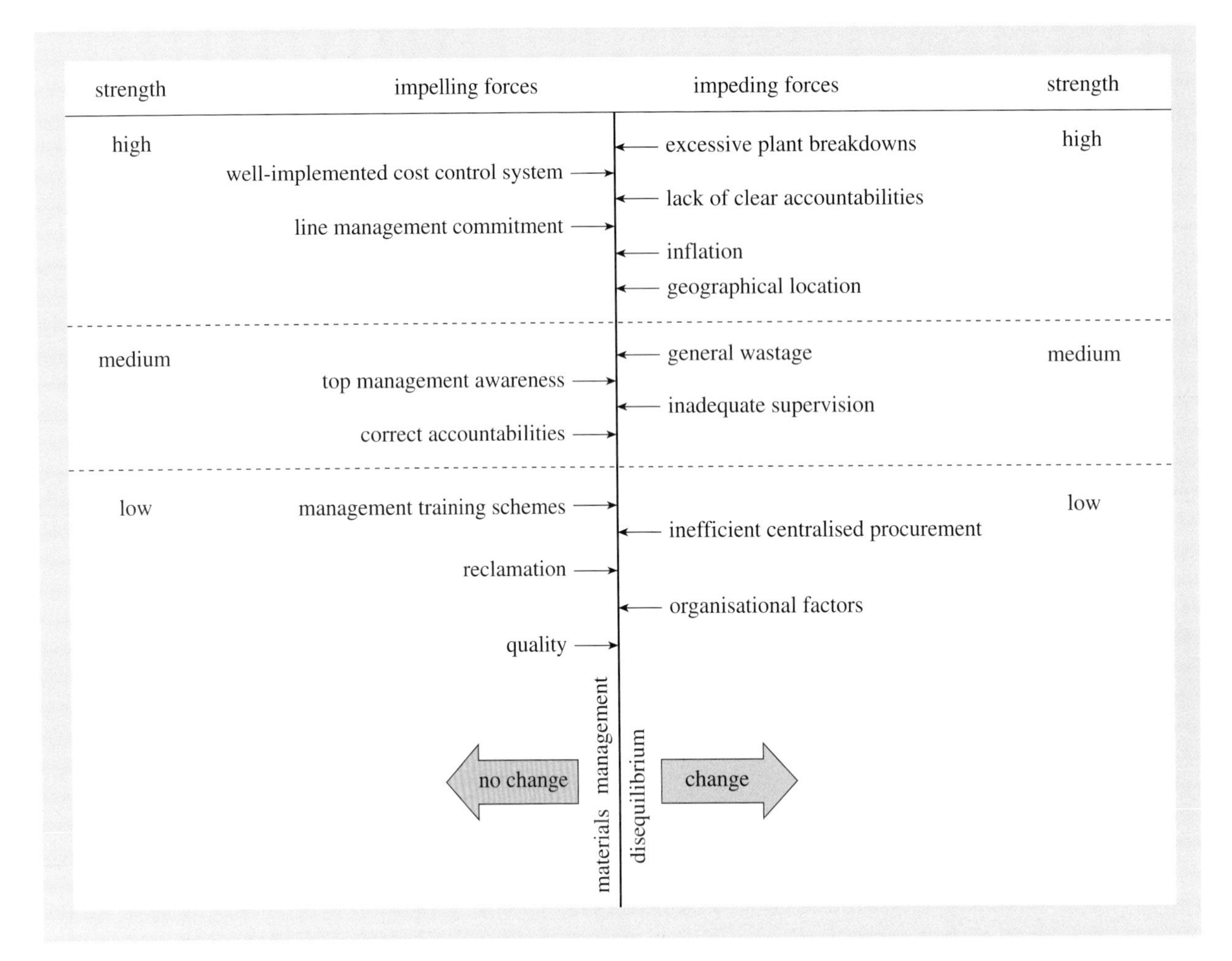

Figure 3.24 Force field analysis – format 2

development in agriculture, predominantly by two firms, Zeneca and Novartis. Figure 3.26 combines company perspectives on sustainable development and commercial viability and also takes into account views from the North American based company Monsanto.

(a) What are the heads and tails shown on Figures 3.25 and 3.26?

(b) Using the construct, 'use the sun's energy to give pest control ... use oil and energy and produce emissions', write a sentence that expands the meaning of this construct.

(c) For the linked constructs in Figure 3.25, 'use less pesticide' and 'build a factory', write a sentence that describes the relationship. ●

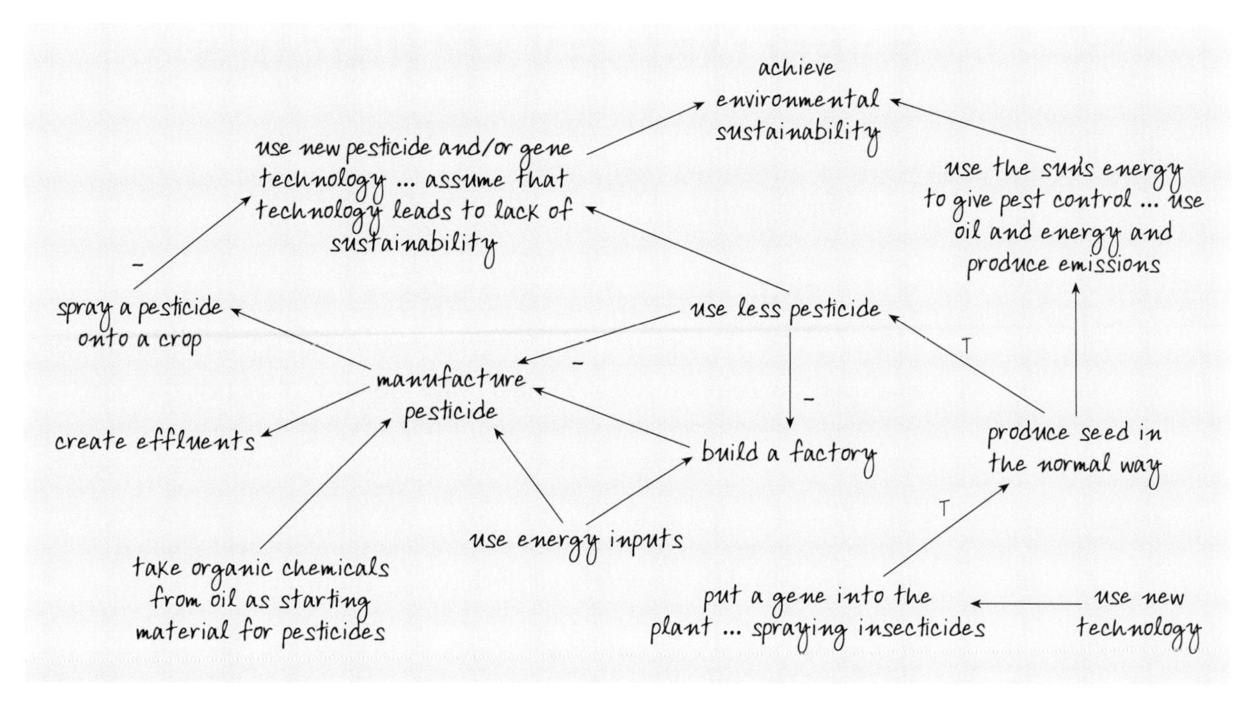

Figure 3.25 Cognitive map 1 (Source: Tait and Chataway, 2003, Figure 1)

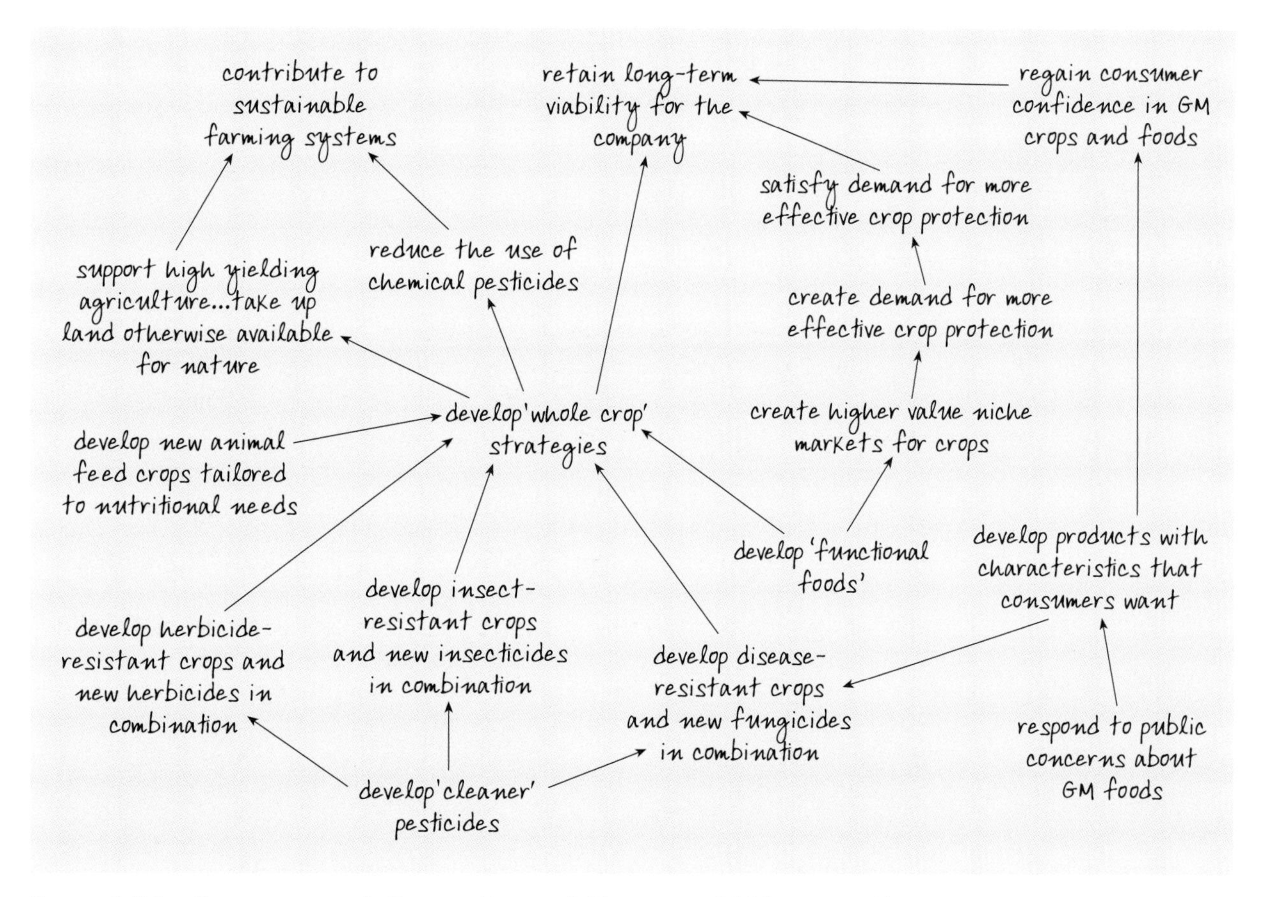

Figure 3.26 Cognitive map 2 (Source: Tait and Chataway, 2003, Figure 2)

7 RELIABILITY TECHNIQUES

Two techniques will be presented in this section:

- failure modes and effects analysis
- fault tree analysis.

7.1 Failure modes and effects analysis

Failure modes and effects analysis (FMEA) is a systematic way of assuring that every conceivable potential failure of a proposed or an actual product, process or system has been considered. British Standard advice on this technique is available as BS EN 60812:2006 *Analysis techniques for system reliability – procedure for failure mode and effects analysis*.

The object of using FMEA is to indicate the key areas where action is necessary to reduce the likelihood of failure. It is often portrayed as a technique that is used only during the design of products, processes or systems but that is definitely not the case. FMEA can also be a valuable aid to improvement.

There are a number of ways of presenting the results of an FMEA. Table 3.5 shows one option, using the example of a ballpoint pen. Table 3.6 is an extract from a table that has been adapted for use when looking at a process (note that RPN is risk priority number).

Table 3.5 FMEA table

Part[a]	Function	Potential failure mode	Potential effects of failure	Severity	Potential causes of failure	Occurrence	How will the potential failure be detected?	Detection	RPN[b]	Actions
Outer tube	Provide grip for writer	Hole gets blocked	Vacuum on ink supply stops flow	7	Debris ingress into hole	3	Check clearance of hole	5	105	Make hole larger Remove cap
Ink	Provide writing medium	Incorrect viscosity	High flow	4	Too much solvent	2	QC on ink supply	4	32	Introduce more rigid QC
Ink	Provide writing medium	Incorrect viscosity	Low flow	4	Too little solvent	2	QC on ink supply	3	24	No action required
Ball and seat	Meter ink supply	Incorrect fit	Ball detached	8	Total failure	2	Inspection checks	2	32	
Ball and seat	Meter ink supply	Incorrect fit	Ball loose	6	Blotchy writing	3	Sampling checks	6	108	Introduce in-process checks during manufacture
Ball and seat	Meter ink supply	Incorrect fit	Ball tight	7	Intermittent writing	4	Sampling checks	6	168	Introduce in-process checks during manufacture Control ball and seat variation
Inner tube	Contain ink	Tube kinked	Ink flow restricted	5	Poor handling in manufacturing	2	No current checks or tests	8	80	Introduce detection checks for this failure
Plug	Close outer tube end	Wrong size	Cannot be fitted	2	Moulding process not in control	2	During assembly	1	4	No action required
Plug	Close outer tube end	Wrong size	Falls out	4	Moulding process not in control	2	No current checks or tests	8	64	Eliminate part Control part moulding process variations

[a] You may see 'part' referred to as 'element' in other texts.

[b] RPN stands for risk priority number.

(Source: Fox, 1993, p. 146)

Table 3.6 Extract from an FMEA for legal documents process failures

Process step	Potential failure mode	Potential effects of failure	Severity	Potential causes of failure	Occurrence	Current controls	Detection	RPN[a]
Pull sheet creation	Inaccurate entries	Wrong documents pulled	9	Clerk mistake, inattention	8	QC	8	576
Priv. log check	Priv. docs not identified	Wrong documents pulled	8	Clerk mistake, inattention	8	QC	8	512
Document review	Miscounts, wrong documents	Wrong documents produced	8	Clerk mistake, inattention	7	QC	6	336
Tracking form creation	Inaccurate entries	Wrong documents tracked	7	Clerk mistake, inattention	5	QC	6	210

(Source: Snee and Hoerl, 2005, p. 262)

[a] RPN stands for risk priority number.

Most of the options use a selection from the following headings or their equivalents.

Element. The subject of the analysis is broken down into named elements or items. The breakdown should normally be to the lowest level of system description available at the time the FMEA is conducted, but for some specialist applications, such as electronic or control systems using integral modular units as building blocks, the modules rather than their individual components may be the subject of the lowest level of analysis.

Function. The function performed by each element is identified.

Potential failure mode. This covers every way in which each element *could* fail to meet each aspect of its specification and should include random and degradation failures. The question to be asked is 'How could it fail?' not 'Will it fail?'.

Potential effects of failure. This looks at the consequences of failure for each element, together with any dependent failures or secondary or side effects.

Severity of failure. A scale is normally used to express the consequences of failure. In Tables 3.5 and 3.6, 1 indicates a minor nuisance and 10 a very serious consequence. An alternative is to categorise the consequences. An example of a severity classification is shown in Table 3.7.

Table 3.7 Illustrative example of a severity classification for end effects

Class	Severity level	Consequence to persons or environment
IV	Catastrophic	A failure mode which could potentially result in the failure of the system's primary functions and therefore causes serious damage to the system and its environment and/or personal injury.
III	Critical	A failure mode which could potentially result in the failure of the system's primary functions and therefore causes considerable damage to the system and its environment, but which does not constitute a serious threat to life or injury.
II	Marginal	A failure mode, which could potentially degrade system performance function(s) without appreciable damage to system or threat to life or injury.
I	Insignificant	A failure mode which could potentially degrade the system's functions but will cause no damage to the system and does not constitute a threat to life or injury.

(Source BS EN 60812:2006, p. 17)

Potential causes of failure. All mechanisms of failure that could result in each failure mode are listed.

Likelihood of occurrence. The likelihood of each failure mode is usually expressed as a probability or as a score on an evaluation scale. In Tables 3.5 and 3.6 a scale of 1 to 10 has been used with 1 indicating a very low probability of occurrence and 10 a near certainty.

How will the potential failure be detected? Some failures are obvious to the person using the subject of the FMEA, but if this is not the case the means by which failures can be discovered should be listed, for example using external test equipment, carrying out periodic performance checks.

Likelihood of detection. The likelihood of detection can also be expressed as a probability or as a score on a scale. In Tables 3.5 and 3.6 a score of 1 indicates a very high probability of detection and 10 a very low probability.

In Tables 3.5 and 3.6 risk priority numbers (RPN) have also been calculated. The RPN is the product of the likelihood of occurrence, the severity and the likelihood of detection, and can be used to indicate which failure modes should be given priority. The actions to be taken, based on the findings of the analysis, are also included.

ACTIVITY 3.10

Imagine you are carrying out an FMEA on a car. List eight symptoms of failure that the driver might notice or experience. ●

A variation of the FMEA principle is the safety analysis table illustrated in Table 3.8. The main difference is that in this case hazards rather than components are listed. This is useful when only a rough idea exists of how the product, service or process will be constructed and it is therefore not possible to draw up a detailed list of elements. The example in Table 3.8 relates to electrical system failures in a missile.

Table 3.8 One form of safety analysis table – electrical system failures

Ref. no.	Condition, event or cause	Relative frequency	Preventive or corrective measures
1	Battery failure	Chronic	There is only one battery in the missile for operational purposes. (A few are provided with instrumentation batteries.) It therefore constitutes a component through which a 'single-point' failure could cause loss of a missile. If redundancy is impracticable, the battery must be an extremely high reliability item. It should be considered a critical item.
1.1	Overheating	Remote	Use current limiters to ensure any possibility of excessive current drain over extended periods is avoided. Ensure battery capacity is adequate to carry any load under all foreseeable conditions.
1.1.1	Excessive current drain		
1.1.2	High internal resistance		
1.2	Bursting	Random	Minimize any type of overheating which could cause gas or electrolyte expansions in case. Maintain battery within prescribed temperature limits. Procedures for assembly and before installation of battery should ensure relief valve operates properly.
1.2.1	Overpressurization		
1.2.2	Plugged or faulty relief valve		
1.2.3	Defective manufacture of container		Maintain close quality control of cases and relief valves.
1.2.4	Hydrogen produced by cell reversal pressurizes container or ignites		Preventing excessive drain will prevent cell reversal and generation of hydrogen gas.
1.3	Lack or leakage of electrolyte	Chronic	Ensure cell container is not cracked before filling. Ensure reservoir is full. Handle with care to avoid container damage. Ensure fill caps are secure after filling. Avoid incompatibility of materials which could cause corrosion and lead to leakage. Ensure brazed joints in manifolds and other parts are leaktight. Use helium or dye leak tests to ensure overall tightness. Investigate machine methods of brazing, since present manual methods could result in wide variations and poor quality products.
1.3.1	Incompatibility and attack of seals, joints, diaphragms, and other components		
1.3.2	Deterioration of battery case with age		

Ref. no.	Condition, event or cause	Relative frequency	Preventive or corrective measures
1.3.3	Cracking: minor fault developed into a crack due to vibration or flight load		
1.3.4	Joints not leak tight		
1.4	Terminal or connection failure	Chronic	Inspect during installation to ensure connections are made properly. Ensure no physical load is on connection. Ensure connector halves are clean, contain no debris, corrosion, solder, cut wire, or dirt. Seal to prevent entrance of moisture, KOH, or other foreign material. Make connections tight to prevent loosening by vibration. Ensure pins are wired correctly by having individual pin assemblies checked visually, and the entire connector tested electrically. Inspect for bent pins prior to assembly. No undue force to be used to mate the halves of the connector.
1.4.1	Connection improperly made		
1.4.2	Physical load causes separation		
1.4.3	Vibration causes separation		
1.4.4	Foreign matter in connector halves		
1.4.5	Moisture or KOH [potassium hydroxide] in connector		
1.4.6	Improper pin wiring		

(Source: Hammer, 1972, p. 177)

7.2 Fault tree analysis

Fault tree analysis involves recording failures that can contribute to an undesired event and representing them diagrammatically. Each failure is analysed in terms of its possible direct causes and then the origins of those causes are looked for. It is then possible to find ways of avoiding the origins and causes. As its name suggests, the output of this type of analysis is a tree-like structure of the form shown in Figure 3.27, so it is in fact a form of the third new tool, tree diagrams, that you looked at in Section 3.3.

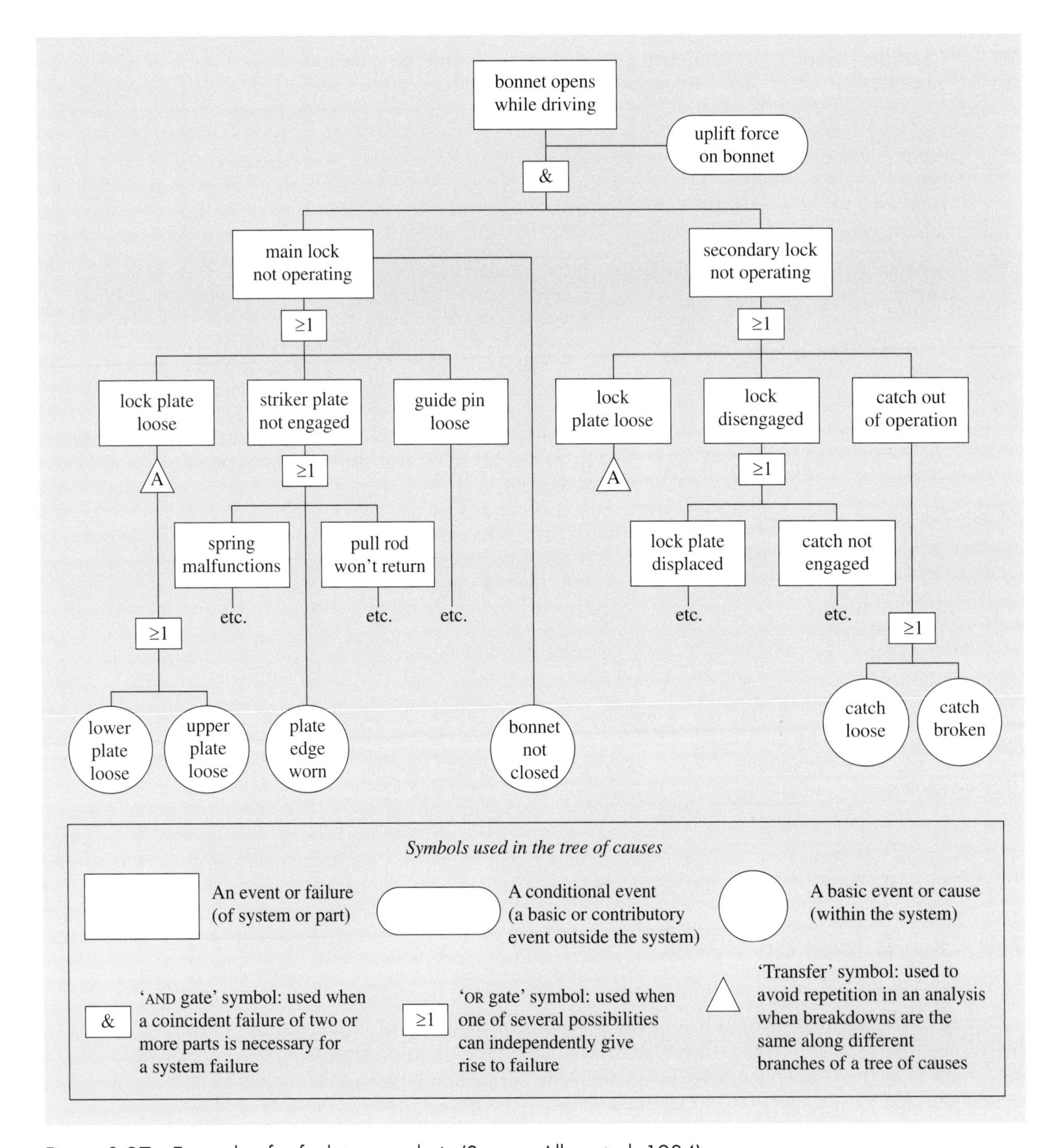

Figure 3.27 Example of a fault tree analysis (Source: Allen et al.,1984)

BS 5760-7:1991 *Reliability of systems, equipment and components – Part 7: Guide to fault tree analysis* defines fault tree analysis thus:

> Fault tree analysis is concerned with the identification and analysis of conditions and factors which cause or contribute to the occurrence of a defined undesirable event, usually one which significantly affects system performance, economy, safety or other required characteristics.
>
> (BS 5760-7:1991, p. 1)

The basic elements of a fault tree are the same regardless of the types of events being analysed. The symbols used in its construction, some of which can be seen in the sample fault tree in Figure 3.27, are as follows:

Circle. The circle represents a basic fault event.

Diamond. The diamond represents a fault event that is assumed to be basic in a given fault tree. It could be divided further to show how it is caused from more basic failures, but this is not done because of lack of time or information or because it is not thought significant enough.

Rectangle. The rectangle represents an event that results from the combination of events of the types described above.

AND *gate*. The output event after an AND gate occurs if and only if all the input events occur.

OR *gate*. The output event after an OR gate occurs if one or more of the input events occur.

Transfer gates. The transfer symbol avoids the need to repeat sections of the fault tree. It provides a reference to another part of the fault tree where an identical sequence of events could be repeated.

Oval. The oval represents a conditional event (a basic or contributory event outside the scope of the analysis).

The first step in constructing a tree is to define a single 'top event'. This will be the starting point for the tree and it needs to be defined in a way that leaves no room for individual interpretation. The next stage is to use knowledge and experience of what can happen to identify the events that lead directly to the top event and present them as the main branches on the tree, connected by the appropriate logic gates. Each branch then needs to be populated with sub-branches and sub-sub-branches and so on until a basic event (one which cannot be subdivided) is reached or until a transfer gate can be used to indicate that the event also appears elsewhere on the tree.

8 CREATIVITY AND IDEA GENERATION

Idea generation is necessary for almost all improvement activities and plays an extremely important part in very many of the methods and techniques covered in this block. Typically, problems have multiple causes and the knowledge required to investigate and solve them lies in the minds of a number of different people. It is therefore essential that mechanisms are available to allow that knowledge to be tapped. The same mechanisms can also be used to enhance creativity. An interesting (and somewhat controversial) call for creativity in the context of improvement has been made by Ackoff:

> Continuous improvement implies small incremental changes made close together in time. This precludes creative leaps, quantum improvements. The creative act necessarily produces discontinuities, qualitative changes. Creativity involves three steps: (1) identification of a self-imposed constraint (assumption), (2) removing it, and (3) exploring the consequences of the removal. The removal of the constraint produces a discontinuity. This is why there is always an element of surprise when we are first exposed to a creative work: it embodies the denial of something we have (usually unconsciously) taken for granted. ...
>
> The point is that *creative but discontinuous improvements* may be worth much more than a string of *small but continuous improvements*. ...
>
> Continuous improvement is likely to enable one only to *keep up*, but not to *take the lead*. It takes creativity to take the lead.
>
> (Ackoff, 1991, pp. 543, 544)

A vast array of creativity and idea generation techniques is available. The Appendix shows the very comprehensive list of techniques studied by students of The Open University Business School course, B822 *Creativity, innovation and change*. In this course I shall examine just a few of those available:

- brainstorming
- brainwriting
- nominal group technique
- SCAMPER
- creative problem solving
- six hats and lateral thinking.

All these techniques are based on principles such as association, abstraction, combination, variation, and the transfer of structures between unconnected areas. Some of these principles help to generate ideas intuitively (association and transfer of structures) while others (variation, combination and abstraction) encourage a more systematic and analytical approach. Between them they cover the three types of idea-generating mechanisms identified by:

> 1 Combining: Grouping known and used ideas in an unusual manner. This is the most widely used mechanism in most activities.
>
> 2 Exploring: This is a progressive change in the conceptual structure through systematic application of known transformations, but used in an unusual manner.
>
> 3 Transforming: this is the creation of a new framework, a different way of thinking.
>
> (Boden, 1994, quoted in Muñoz-Seca and Riverola, 2004, p. 244)

8.1 Brainstorming

Although idea generation occupies such a central and important position in problem solving and improvement, many organisations rely on just one technique: brainstorming. It can be argued, however, that brainstorming is not always the most appropriate idea generation technique, that it suffers from a number of drawbacks, and that those drawbacks are frequently overlooked to the detriment of the quality of the ideas that are generated. At the same time that brainstorming is being used ineffectively, other techniques, which could provide greater insight into problems and generate better solutions, are being neglected.

On the face of it, brainstorming is a simple method for generating ideas and it is likely that this apparent simplicity is the main reason for its popularity. In reality, however, the rules of brainstorming are far more rigorous than many users appreciate. The following rules should typically be followed by participants:

1. Work in small, close groups of perhaps five or six, in a private, protected area away from interference or interruption.
2. Create an atmosphere that is safe, supportive, concerned with encouragement and building (not criticism or dissection), fun, playful, energetic, enthusiastic, permissive, stimulating and risk taking.
3. Divide the time into periods of relaxed privacy for individual imagination and contrasting periods of excited, lively, rapid-fire group interaction.
4. Write up all the ideas as they occur where everyone can see them.
5. Treat everyone as equal and enable everyone to contribute, although it may be useful to appoint a 'compère' who discourages criticism, encourages dramatic and outlandish ideas, and maintains the pace.

6 Continue while the excitement lasts, but stop at the first signs of staleness.
7 When a good stock of raw ideas has been generated, switch into a more controlled mode and work through each idea in turn, asking: 'How could we develop this idea to make it into a usable option?' Ideas that at first sight seem particularly silly may well turn out to be the most useful after this secondary analysis.

ACTIVITY 3.11

Which three of the seven rules listed above would you select as being the most important to the success of a brainstorming session? ●

In practice, it is very difficult to ensure that these rules are followed. The stifling of critical and negative remarks, for example, requires a good deal of self-discipline. 'Everyone is equal and everyone contributes' may not have been a problem in Japanese-style quality circles, where the members are of approximately equal status. However, tensions arising from hierarchical differences can cause difficulties in the UK, where many organisations have problem-solving and improvement teams that incorporate managers, experts and shop-floor workers in the same team. Although the members may be notionally equal in so far as their problem-solving and improvement activities are concerned, they must sometimes find it difficult to forget the differences of status that prevail during the rest of the working week. A similar problem concerns the level of each individual's knowledge about the situation under consideration. There is an optimum level of knowledge: the expert may know too much, especially in relation to what is not possible, while others may know too little.

In some organisations, members of problem-solving and improvement teams are self-selected. While this self-selection can be highly beneficial in terms of the motivation of members and the credibility given to their output, it can impose difficulties for brainstorming. The technique does not incorporate any mechanisms for dealing with what can best be described as problem people: the very talkative, the know-it-all, the permanent critic, the compulsive joke-maker and the extremely shy.

Another drawback of brainstorming is its extreme familiarity. When the use of a creativity technique becomes routine, the sorts of idea that are generated can also become routine. The use of different methods, on the other hand, can introduce variety and make the idea generation process more interesting and hence more productive.

Brainstorming should be about generating ideas, not about deciding which ideas are feasible, but all too often the term is used to describe a meeting at which people organise a list of suggestions that exist already in their minds. Ideas are thus put forward rather than generated. Furthermore, the session is often terminated when a solution that seems feasible is presented, rather than being continued until all possibilities have been exhausted. In fairness, however, if the rules of brainstorming are observed the technique is well able to fulfil its aim, which is to provide conditions that stimulate and support divergent thinking and generate novel or creative ideas that dissolve problems or provide ingenious solutions.

8.2 Brainwriting

Brainwriting is most useful at the level of the individual problem. It is close to brainstorming in that it too is based on association. However, it differs from brainstorming in one major respect: the participants do not talk to each other; they communicate in writing or even by electronic means. Some of the ways of organising a brainwriting session are outlined below.

The pool method

Each participant writes four ideas on a blank sheet of paper and puts it in a pool at the centre of the table. Each then takes further sheets of blank paper, writes four ideas on each sheet and places them in the pool until no more ideas are forthcoming. At this point each participant takes from the pool a sheet that has been written by someone else and tries to add to or develop the ideas on it, being careful not just to repeat ideas that they have already presented on previous sheets. This process of working with other team members' ideas continues until the end of the session.

The card-circulating technique

Piles of blank cards are placed between adjacent members of the team. Each time that a participant has an idea they write it on a blank card from their left and then place the card on their right. When a participant feels in need of external stimulation, they look through the cards in the pile on their left that other participants have written on and try to develop one or more of the ideas using association. When this has been done, the participant places each of these cards too on the pile on their right, and so on. After a session lasting in the region of 25 minutes, ideas with similar themes are grouped together for further consideration.

The 635 method

This method sounds a little like a party game! Six people each write down three ideas in five minutes on a form. The forms are passed on five times. After every pass each participant reads the ideas on the form that they have just received and tries to develop them further through association.

The collective notebook method

This method is different from the previous ones in that the participants do not need to meet as a team while they are carrying it out. It is therefore very useful for shift workers or for people working at a number of different sites or working at home. Because of the nature of the method, it is also particularly useful for problems where observation of processes or operations over periods of time is necessary in order to detect weaknesses or unfulfilled requirements. Eight to ten people are each asked to record at least one idea per day in a notebook for one week. At the end of the week, the notebooks are redistributed among the participants and each tries to build on the ideas in the notebook received, again working on at least one idea per day. The notebooks are redistributed again after three or four days and so on for a total period of about four weeks.

The electronic method

Through the use of networks and computer-conferencing facilities it is possible to adapt the pool method and the card-circulating technique so that they can also be used by people who are working at a distance from each other.

One of the advantages of brainwriting is the variety of methods that it offers. By switching between them it is possible to diminish the problems of over-familiarity that can grow when a small group of people tries to use the same technique over and over again.

8.3 Nominal group technique

The nominal group technique (NGT) developed out of psychological research by Delbecq and Van De Ven at the University of Wisconsin and has been in use since 1968. Its special features are participation (everyone is encouraged to join in) and democratisation (the voice of the lay person is strengthened against possibly domineering professionals). It thus specifically addresses some of the deficiencies of brainstorming that I identified in Section 8.1.

Essentially, the NGT aims to ensure that the full range of views and ideas of a group is heard and considered, and that the plan or agenda that emerges truly reflects the members' concerns and is supported by them. It is therefore particularly useful when seeking solutions to higher-level problems, such as the identification of goals, strategic planning and preparation for later work. This is especially so when controversial problems and decisions are being considered, for example, how to design and implement an auditing system or whether to move from a top-down to a bottom-up approach to problem solving. A further reason why the technique is very useful in tackling higher-level problems is that it combines divergent and convergent thinking, whereas both brainstorming and brainwriting end after the divergent phase.

There is no leader or chairperson at an NGT meeting, merely someone who acts as recorder. The process used involves four steps:

1. silent and individual generation of ideas in writing
2. round-robin recording of ideas as simple phrases on a flip chart

3 discussion to clarify the ideas put forward
4 voting on a priority of ideas, where the group decision is derived by rank ordering or rating.

Typically, two or three hours are needed for a full NGT meeting. This is significantly longer than the span of time usually spent on a brainstorming session. Indeed, it is a much longer time than that taken for the whole of a typical improvement team meeting. However, the higher-level problems for which the NGT is particularly appropriate are usually addressed much less frequently than individual problems.

8.4 SCAMPER

SCAMPER is based on a checklist designed to generate suggestions for changing existing ideas or products into something new. The name SCAMPER comes from the initial letters of the items on the checklist:

> Substitute: components, materials, people, places.
>
> Combine: mix, blend, combine with other parts or services, integrate.
>
> Adapt: alter, change function, use part of another element.
>
> Modify: increase or reduce in scale, change shape, modify properties (e.g. colour).
>
> Put to another use or use in another place, find a different market.
>
> Eliminate: remove elements, simplify, streamline, reduce to core functionality.
>
> Reverse: turn inside out or upside down.

The following is an example from a producer of computers and printers looking for new products:

> Substitute – use of high tech materials for specific markets – use high-speed components?
>
> Combine – integrate computer and printer, printer and scanner
>
> Adapt – put high quality ink in printer, use high quality paper
>
> Modify – produce different shape, size and design of printer and computer
>
> Put to another use – printers as photocopies or fax machines
>
> Eliminate – eliminate speakers, colour screens, colour ink etc.
>
> Reverse – make computer desks as well as computers and printers, or computer chairs etc.
>
> (Mycoted, 2006)

8.5 Creative problem solving

Creative problem solving (CPS) was developed in the 1950s by Sidney J. Parnes. Figure 3.28 shows an overview of the process and Box 3.3 gives a more detailed outline.

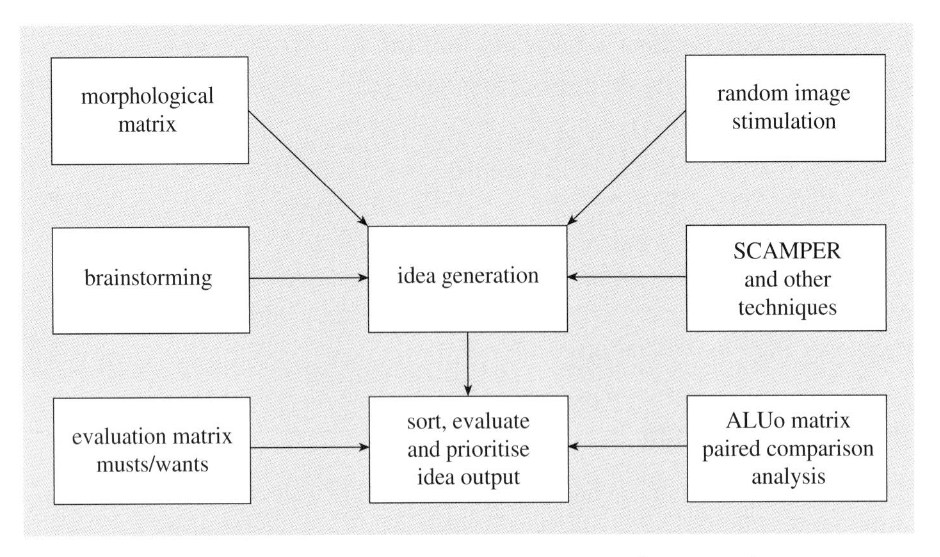

Figure 3.28 The basic CPS process (Source: redrawn from Hipple, 2005, p. 25)

BOX 3.3 THE '6×2 STAGES' FORM OF CREATIVE PROBLEM SOLVING

1 *Mess finding* Sensitize yourself (scan, search) for issues (concerns, challenges, opportunities, etc.) that need to be tackled.

   Divergent techniques include 'Wouldn't it be nice if ...' (WIBNI) and 'Wouldn't it be awful if ...' (WIBAI), i.e. brainstorming to identify desirable outcomes, and obstacles to be overcome.

   Convergent techniques include the identification of hotspots (*Highlighting*), expressed as a list of IWWMs ('In what ways might ...'), and selection in terms of ownership criteria (e.g. problem-owner's motivation and ability to influence it) and outlook criteria (e.g. urgency, familiarity, stability).

2 *Data finding* Gather information about the problem.

   Divergent techniques include *Five Ws and H* (who, why, what, when, where and how) and listing of wants, sources and data. List all your information 'wants' as a series of questions; for each, list possible sources of answers; then follow these up and for each source, list what you found.

   Convergent techniques again include: identifying hotspots (*Highlighting*); *Mind-mapping* to sort and classify the information gathered; and also restating the problem in the light of your richer understanding of it.

3 *Problem finding.* Convert a fuzzy statement of the problem into a broad statement more suitable for idea finding.

   Divergent techniques include asking 'Why?' ... .

   Convergent techniques include *Highlighting* so that the statements contain only one problem and no criteria, and selection of the most promising statement (but note that the mental 'stretching' that the activity gives to the participants can be as important as the actual statement chosen).

4 *Idea finding.* Generate as many ideas as possible.

Divergence can use any of a very wide range of idea-generating techniques. The general rules of classical brainstorming (such as deferring judgement) are likely to underpin all of these.

Convergence can again involve hotspots or mind-mapping, the combining of different ideas, and the shortlisting of the most promising handful, perhaps with some thought for the more obvious evaluation criteria, but not over-restrictively.

5 *Solution finding* Generate and select clear evaluation criteria (using an expansion/contraction cycle) and improve (which may include combining) the shortlisted ideas from 'Idea finding' as much as you can in the light of these criteria. Then select the best of these improved ideas.

6 *Acceptance finding* How can the idea you have just chosen be made acceptable and implemented? Avoid negativity, and continuc to apply deferred judgement – problems are uncovered to be solved, not to discourage progress. Action plans are better developed in small groups of two or three than in a large group (unless you particularly want commitment by the whole group). Particularly for 'people' problems it is often worth developing several alternative action plans. Possible techniques include *five Ws and H* ... .

(Source: T837 Block 6, Part 5, pp. 4–5).

As you can see, creative problem solving separates the idea generation phase from the idea evaluation process. It also emphasises the need to generate large numbers of ideas, the assumption being that the greater the number of ideas generated, the more likely they will help to solve the problem under consideration. In the divergent phase, techniques such as those you have already met in this section are used to generate ideas. The purpose of the convergent phase is to narrow down and focus the ideas generated in the divergent phase. Hipple (2005) recommends the following tools for this second stage:

ALUo
This focusing tool asks the participants to analyse each idea in terms of its **A**dvantages, **L**imitations, **U**nique qualities, and **o**vercome limitations to assist in idea optimization, prioritization and selection.

> Evaluation matrix
> This tool lists the ideas selected for final evaluation against evaluation criteria established by the problem-solving group or its organization. In a simple table, the ideas are listed vertically and the criteria listed horizontally. At each intersection, a rating or relative ranking is made, assisting in the final decision-making of the group.
>
> [...]
>
> [D]istinguishing between musts and wants ...
>
> [D]eliberate paired [comparison] analysis of ideas against each other.
>
> (Hipple, 2005, p. 26)

8.6 Six hats and lateral thinking

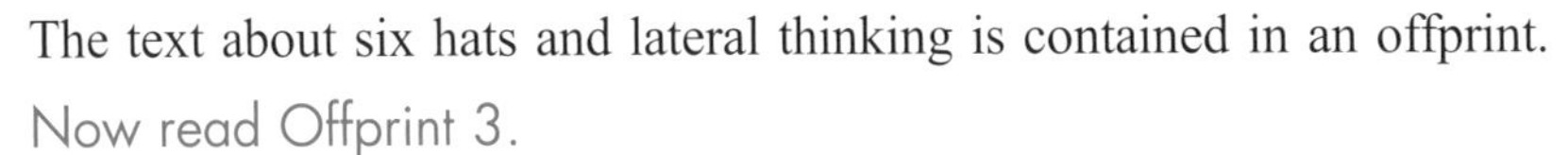

The text about six hats and lateral thinking is contained in an offprint.

Now read Offprint 3.

9 INVESTIGATING CONTEXT AND LOOKING FOR OPPORTUNITIES

The techniques that will be covered in this section are:

- stakeholder analysis
- SWOT analysis
- environmental scanning
- benchmarking
- gap analysis.

9.1 Stakeholder analysis

In problem solving and improvement the notion of stakeholder is extremely powerful. Solutions that ignore the requirements of key players and fail to take into account the views of the wider constituency are doomed to failure. Stakeholder analysis has at least three components:

1. Identify stakeholders.
2. Prioritise stakeholders.
3. Understand key stakeholders.

One tool that is frequently deployed in this analysis is the power/interest grid shown in Figure 3.29. This is, however, a blunt instrument, especially if individual stakeholders are just considered separately. Disparate groups of stakeholders, each with very limited amounts of power individually, can wield a surprising amount of power if they unite in common cause. For this reason, instead of just plotting the positions of individual stakeholders on the grid, it can be very useful to use a technique such as a relationship map or

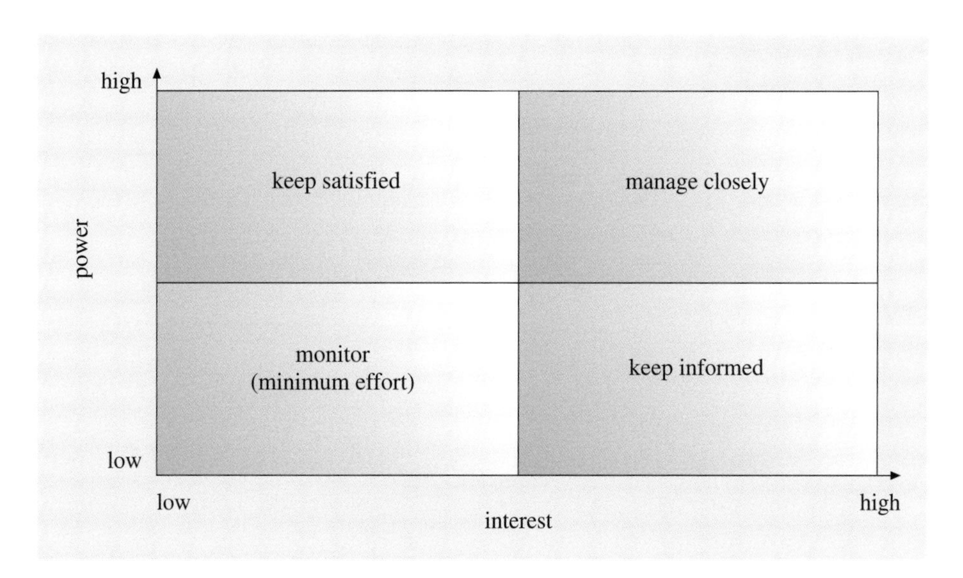

Figure 3.29 Power/interest grid (Source: redrawn from Mind Tools, n.d.)

an affinity diagram to study the relationships (if any) between the stakeholders in individual quadrants.

Now read Offprint 4.

ACTIVITY 3.12

Imagine you are employed by a major supermarket chain and are undertaking a project to increase the effectiveness of food safety measures. Suggest eight stakeholders who would have an interest in your project. ●

9.2 SWOT analysis

SWOT analysis is a strategic planning tool that is used to evaluate the key internal and external factors that are important in trying to reach a predefined objective. The internal factors are divided into strengths and weaknesses, and the external into opportunities and threats; hence the name SWOT. The standard form for presenting the analysis is shown in Figure 3.30. The example deals with health technology assessment (HTA).

strengths	weaknesses
• interest expressed by different actors/players (ministry, university, sickness funds, physicians) • 'initiative group' skilled and committed, with contacts in ministry • international contacts • external support (such as World Bank) • linkage between the first two of the above • well-developed information technology • infrastructure for training • linkage with HTA community (ISTAHC member) • good resource centre (university library) • multidisciplinary team • public access to the internet	• poor communication between stakeholders • lack of awareness of HTA in general • poor data availability • shortage of information specialists in health information • no needs assessment done • insufficient manpower
opportunities	**threats**
• decentralisation • health reform initiated • demand for HTA • limited resources (argument for developing HTA) • bring together multiple stakeholders • academic growth • ongoing development of new technologies • good timing (continuing World Bank Health Project) • demand for more transparent decision-making • development of HTA with a broader social focus • international support	• funding • attitudes of decision-makers/bureaucrats • no broadly accepted priorities in health policy

Figure 3.30 SWOT analysis (Source: Gibis et al., 2001, p. 32)

When using the technique for problem solving and improvement, questions that it is useful to ask include:

How can each strength be used?

How can each weakness be overcome?

How can each opportunity be exploited?

How can each threat be negated?

9.3 Environmental scanning

Environmental scanning, also known as environmental analysis or environmental monitoring, is a method of identifying, collecting, analysing and communicating information about external influences. Where problem solving and improvement are concerned it can be used in a number of ways. It can be used to generate information when investigating situations (for example to identify opportunities and threats for a SWOT analysis) and as a source of potential solutions and improvements.

Figure 3.31 shows the very broad scope of environmental scanning and the many different aspects of the environment that might need to be considered. You can see that the environment has been divided into two layers: an inner layer comprising the competitive and/or market environment; and an outer layer representing the macro-environment. The factors shown in the macro-environment are collectively known as STEEPV factors:

Social

Technological

Economic

Environmental

Political

Values

Note that 'Environmental' in STEEPV refers to the natural environment (weather, pollution, etc.) rather than environment as defined earlier in this block.

ACTIVITY 3.13

How was environment defined earlier in this block? ●

Table 3.9 provides some examples of STEEPV factors and of interactions between individual factors or groups of factors.

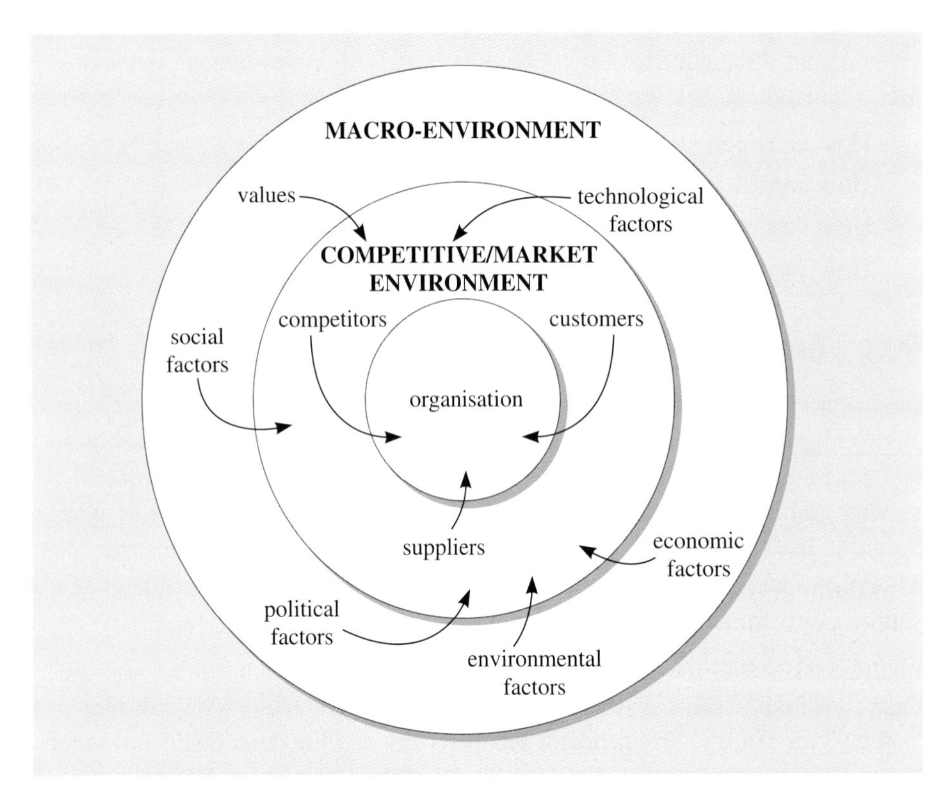

Figure 3.31 The external business environment

Table 3.9 Examples of STEEPV factors in the business environment in 2004

	Examples of factors	**Examples of interactions between individual factors or groups of factors**
Sociological	Demography Trends to smaller households in advanced economies AIDS in Africa	Pressure on pharmaceutical companies to provide anti-AIDS drugs cheaply for third world
Technological	Availability of cheap powerful computing Rapid growth of genomics-based technology (science and applications) Increased uptake of mobile (wireless) communications technology	Increasing tendency to 'global village' model of world, where all countries/people potentially have access to shared information via the internet
Economic	Single European currency Trade with developing countries	Concerns over exploitation of developing countries by multinational organisations seeking to minimise operating costs by locating in poorly regulated regions.

	Examples of factors	Examples of interactions between individual factors or groups of factors
Environmental	Pollution Global warming Biodiversity	Political interventions (legislation) to minimise pollution and global warming
Political	Counter-terrorism as dominant aspect of western governmental defence strategy Environmental legislation	Kyoto agreement to limit carbon dioxide emissions
Values	Work–life balance concerns in advanced economies Ethical corporate governance Environmental awareness	Integration of green issues into mainstream politics

A lot of information for an environmental scan can be collected from colleagues, customers and suppliers. Traditionally, other major sources have been accessed through membership of professional bodies and attendance at events such as trade exhibitions and conferences, but more and more the World Wide Web is becoming a primary source. Figure 3.32 shows one example. It is a picture of the popularity of programming languages as evidenced by book sales. The area of each block shows the current share of the market while the figures show the changes that have occurred compared with the previous year.

9.4 Benchmarking

Benchmarking, also sometimes known as best practice benchmarking (BPB) or competitive benchmarking, began its rise to prominence in the late 1980s and has attracted a lot of interest ever since. It enables an organisation to compare its own activities, processes or performance with those of others. In the earliest examples, organisations compared themselves with others in the same sector but this was soon widened. For instance, South-West Airlines in the USA drew on the methods used by the pit crews at the Indianapolis 500 car race to reduce the turnaround time for its aircraft from 30 to 15 minutes.

Benchmarking is a five-stage procedure as follows.

1 Select area for benchmarking.
2 Decide who to benchmark against:
   - other parts of own company
   - direct competitors
   - parallel industries
   - totally different industries.

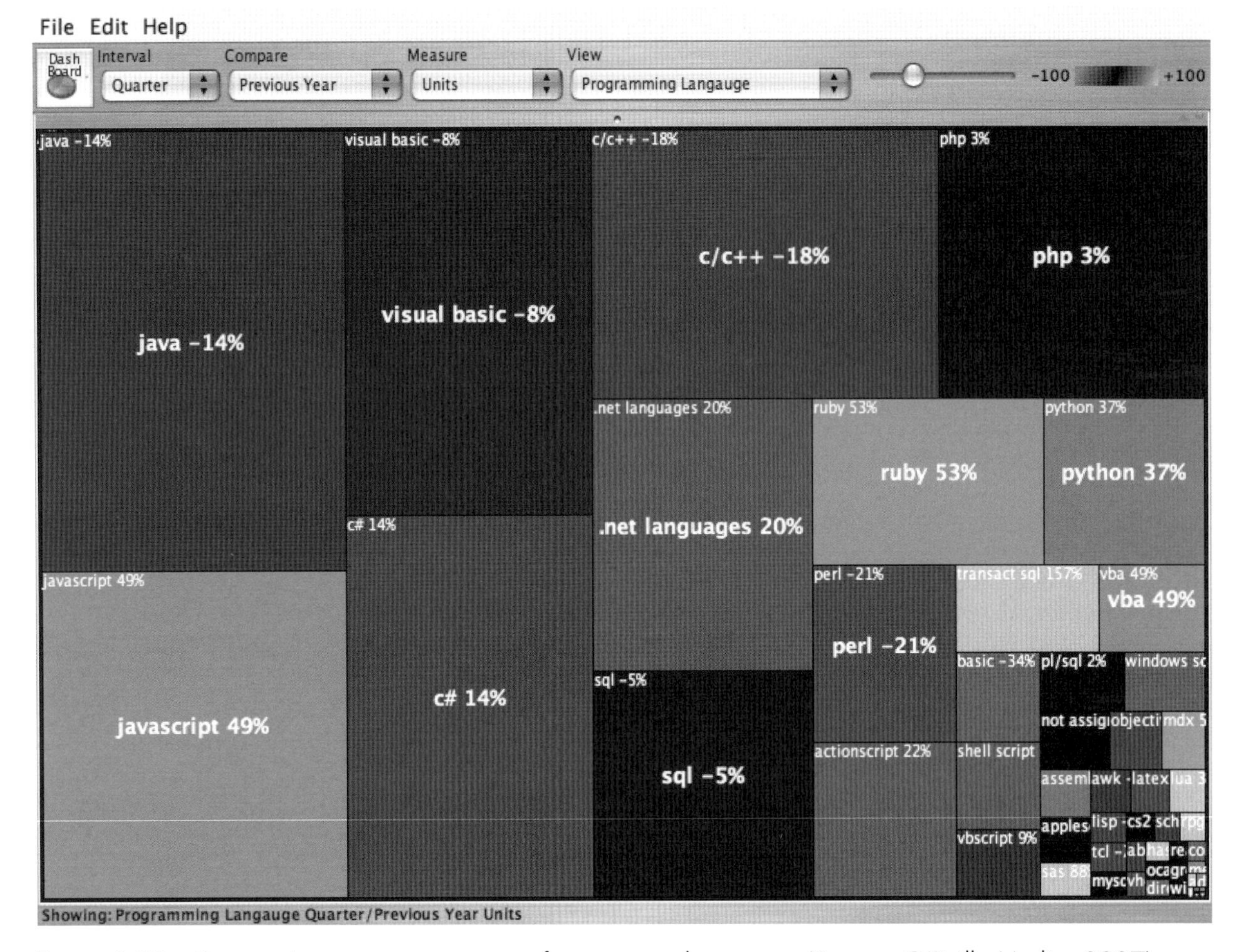

Figure 3.32 Quarter 4 year-on-year treemap for computer languages (Source: O'Reilly Media, 2007)

3 Identify sources of information:
   - published material
   - customers, suppliers and other observers
   - direct exchange.
4 Analyse the information.
5 Use the information.

To date, most organisations have benchmarked against *successes* elsewhere but there is no reason for use of the concept to be restricted in this way. The term 'learning by analogy' has been coined (see Fortune and Peters, 1995) for the valuable insights that could be gained by applying a benchmarking approach to other organisations' failures.

Although benchmarking is almost universally accepted as being an extremely useful technique, it does have some critics. One of them is Ackoff, a leading proponent of systems thinking. His argument runs as follows:

> [I]f we were to combine the best practices relevant to each part of a system – and an organization is a system – we would *not* get the best system. We may not even get a good one. This is apparent from the following example.
>
> Suppose we were to buy one of each type of automobile available in the United States and place them in a large garage. Then we hire a group of the best automotive engineers in the world. We give them the following problem. Which of these cars has the best engine? Suppose they find that the Rolls Royce does. We note this and then ask them to do the same for the transmission. Suppose they find that the Mercedes has the best transmission. We note this and continue until we know which car has the best of each part required for an automobile. When this list is complete we give it to the engineers, ask them to remove the parts listed from their respective automobiles, and assemble them into the best currently possible automobile. This automobile would be made up of the best available parts. But the fact is that we would not even get an automobile, let alone the best one, because *the parts don't fit*. Management should focus on interactions rather than actions. Practices selected by benchmarking seldom take interactions into account. The only type of benchmarking that avoids such an error is benchmarking of systems taken as a whole, systematically, not as aggregations of independent parts.
>
> (Ackoff, 1993, p. 581)

Another author who expresses reservations is Jenkins (2005):

> In practice, the evidence indicates that even in the same company, such lessons as there are, are often hard to learn and that good practices when recognised are not easy to transplant. In terms of learning from stronger competitors, there is again a shortage of success stories. [One example] is the inability of American car makers to learn from their Japanese competitors. They have sought to fathom the hidden secrets of their competitors' success. They initially concluded that the reasons for their rivals' superior productivity was technical advantage. This was why General Motors invested £45 billion in new technology. It soon became apparent that this had done little to make the company more competitive. So they decided that the answer must lie in better styling (that is, in making the car look more attractive). To achieve this aim, huge sums have been invested. It has not made a ha'porth of difference.
>
> The reality is that successful Japanese companies have made no secret of the reasons for their success. There are just two. While American car makers have continued to decide what customers

> can be persuaded to buy, their competitors have sought to find out what their customers want. And while Americans remain hung up on the attractions of 'big bang' – the big idea – their competitors have consistently achieved spectacular improvement by a myriad of incremental innovations.
>
> Every year, the predicament of the American car makers gets worse. The reasons for their rivals' success is transparent, yet the lessons are not learned. All of which lends support to the view that the value of comparisons has been overstated.
>
> (Jenkins, 2005, pp. 34–35)

9.5 Gap analysis

Berry et al. (1985) suggest that, to improve service quality, organisations need to:

- identify primary quality determinants so that they can be the focus for improvement
- manage customer expectations in advance of the service delivery in order to make sure they are realistic
- manage evidence, that is, make sure that the tangibles associated with the service convey the proper clues about its quality
- educate customers about the service so that they are able to make better decisions
- develop a quality culture by establishing specific quality standards, employing people with the capacity to meet those standards, undertaking staff training, verifying that the standards are met, and rewarding success
- automate where appropriate
- follow up the service to identify opportunities for further improvement.

ACTIVITY 3.14

Suggest three tangibles associated with the servicing of domestic heating boilers. ●

In addition to this general advice, Berry et al. have developed a method for investigating and diagnosing deficiencies in quality based on a conceptual model of service quality known as the gap analysis model. This is shown in Figure 3.33.

Gap 1 is the difference between customers' expectations and management's perceptions of those expectations. It is suggested that the size of this information gap depends on the amount of marketing research that is conducted, the amount of upward communication between senior managers and those in contact with customers, and the number of levels of management.

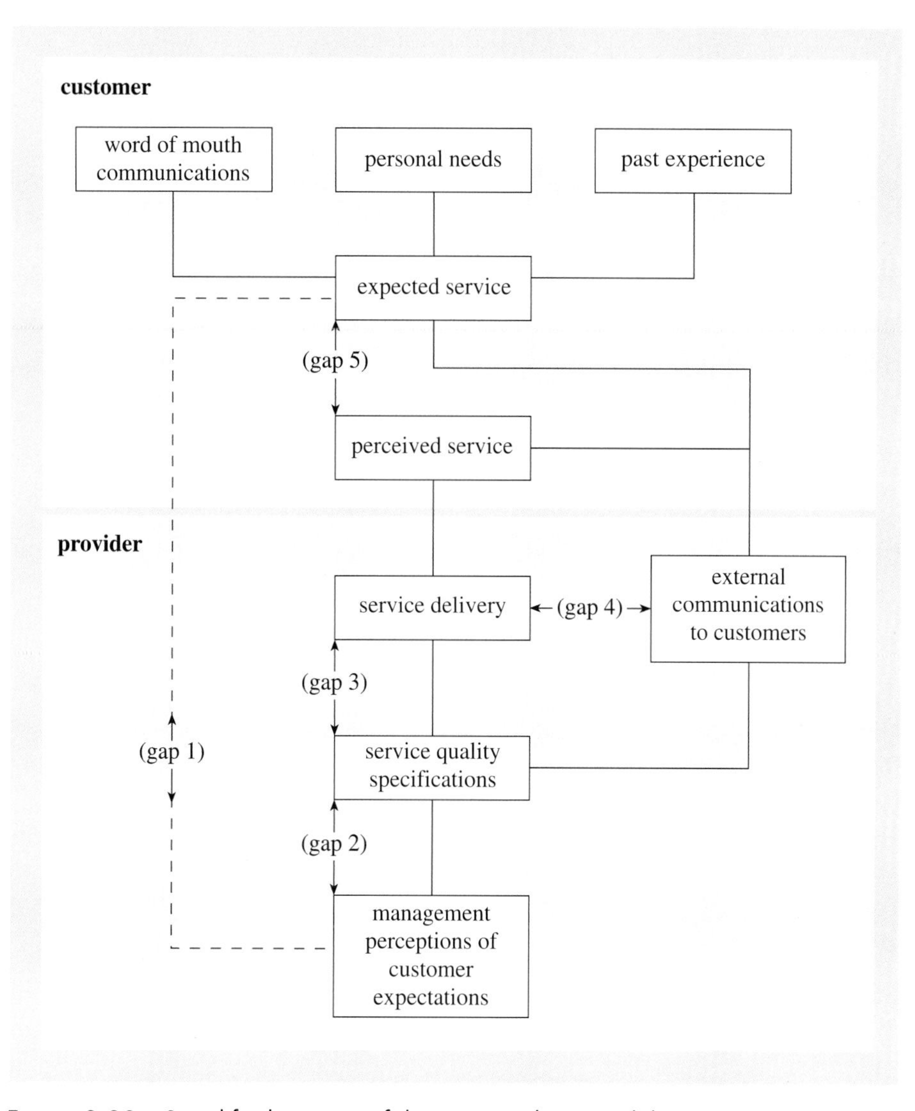

Figure 3.33 Simplified version of the gap analysis model (Source: Parasuraman et al., 1985)

Gap 2 is the difference between management's perceptions of customers' expectations and the actual specifications of the services being offered. It is a planning and design gap that is a function of the level of management commitment to quality, the amount of task standardisation, the extent to which it is regarded as feasible to meet needs, and whether quality goals are set. It can also be due to lack of resources or poor planning or design capabilities.

Gap 3 is the discrepancy between what the service processes are designed to provide and what they actually provide, and it occurs when providers are not doing what they should be doing. In a later paper the size of this gap is linked to: the extent of teamwork perceived by employees; employee–job fit; technology–job fit; the extent of perceived control experienced by customer-contact personnel; the extent to which behavioural control systems are used to supplement output control systems; the extent of role conflict experienced by customer-contact personnel; and the extent of role ambiguity experienced by customer-contact personnel (Zeithaml et al., 1988).

Gap 4 is the difference between what customers are promised and what they receive. It can be ascribed to over-promising and to poor communications between those marketing or publicising a service and those delivering it. The authors cite an unsuccessful advertising campaign by Holiday Inn as an example of this type of gap:

> Holiday Inn's agency used consumer research as the basis for a television campaign promising 'no surprises' to customers. Top managers accepted the campaign in spite of opposition by operations executives who knew that surprises frequently occur in a complex service organization. When the campaign was aired, it raised consumer expectations, gave dissatisfied customers additional grounds on which to vent frustrations, and had to be discontinued.
>
> (Zeithaml et al., p. 44)

Gap 5 is the difference between the quality perceived by the customer and the customer's expected level of service. It is argued that this gap will cease to exist if all of the other four are closed. It can thus provide a way of checking whether quality improvement actions have been effective. Each gap is likely to require a different set of actions to eliminate it.

ACTIVITY 3.15

Suggest separate strategies for closing gaps 1 and 3. ●

10 DESIGN

The techniques covered in this section are:

- poka-yoke
- quality function deployment
- TRIZ
- Y2X
- Taguchi methods
- SMED

The section ends by looking at process redesign.

10.1 Poka-yoke

Poka-yoke or, to give it its more graphic name, 'foolproofing' derives its formal name from two Japanese words: *yoker* (to avoid) and *poka* (inadvertent errors). Its development is usually credited to Shigeo Shingo, who worked on it extensively in the early 1960s, but examples of its application are, in a sense, much older than the technique itself. Look, for example, at a standard British 3-pin plug. Why do you always plug it in the right way up? Because that is the only way it will go! That is foolproofing or mistake proofing.

A classical example of *poka-yoke* can be seen in Figure 3.34. It shows the 'before' and 'after' representations of the positioning of mounting holes on speaker boxes. Before the application of *poka-yoke* it was difficult to determine the correct orientation because the top and bottom holes were symmetrical; now it is simple. Figure 3.35 shows another example. Before *poka-yoke* the design of an integrated circuit was such that it could be inserted backwards. By making the changes shown it was possible to prevent the problem from occurring.

There is, however, slightly more to the technique than this. *Poka-yoke* can be used to provide devices that warn that a mistake is about to be made or has just been made. The prevention or warning is accomplished in one of the following three different ways:

1. *By shutting down the process*. For example, an automatic check-weighing machine can be used to stop a canning line if an under-weight fill is detected.
2. *By controlling the process more carefully*. For example, Nikkan cites the case of a lens-rinsing process which resulted in a number of scratched lenses (Nikkan Kogyo Shimbun Ltd, 1988, p. 174). The problem was caused by the lenses knocking against the blades of the rack that was holding them. By redesigning the rack so that the lenses were held at an angle rather than vertically it was possible to prevent the damage.

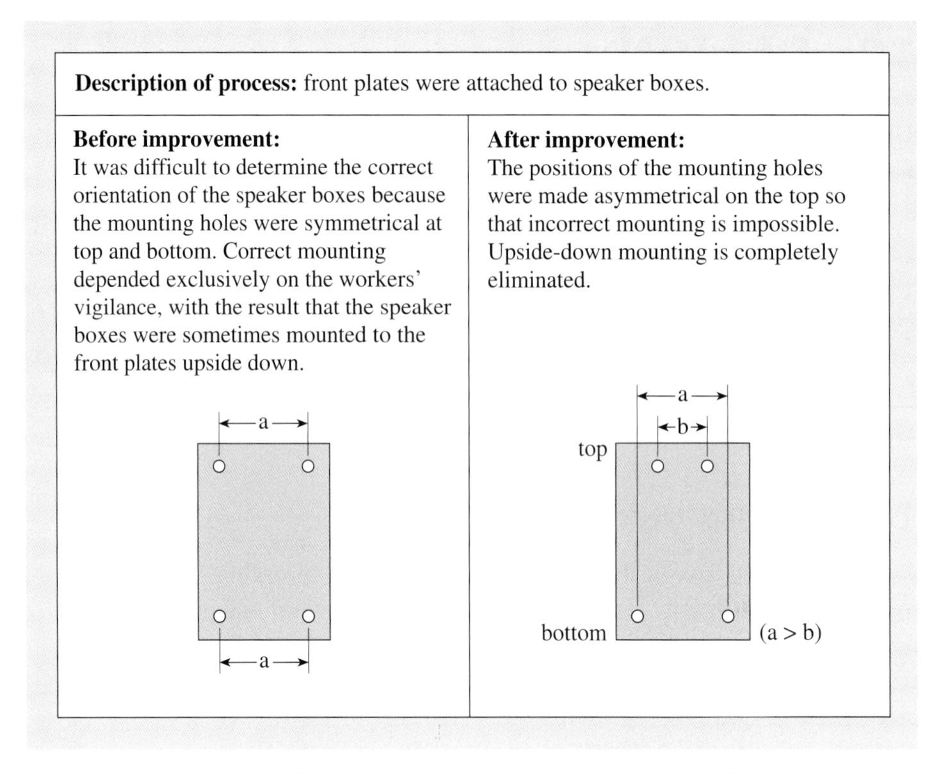

Figure 3.34 A poka-yoke exercise on speaker box assembly (Source: Nikkan Kogyo Shimbun/Factory Magazine, 1988, p. 131)

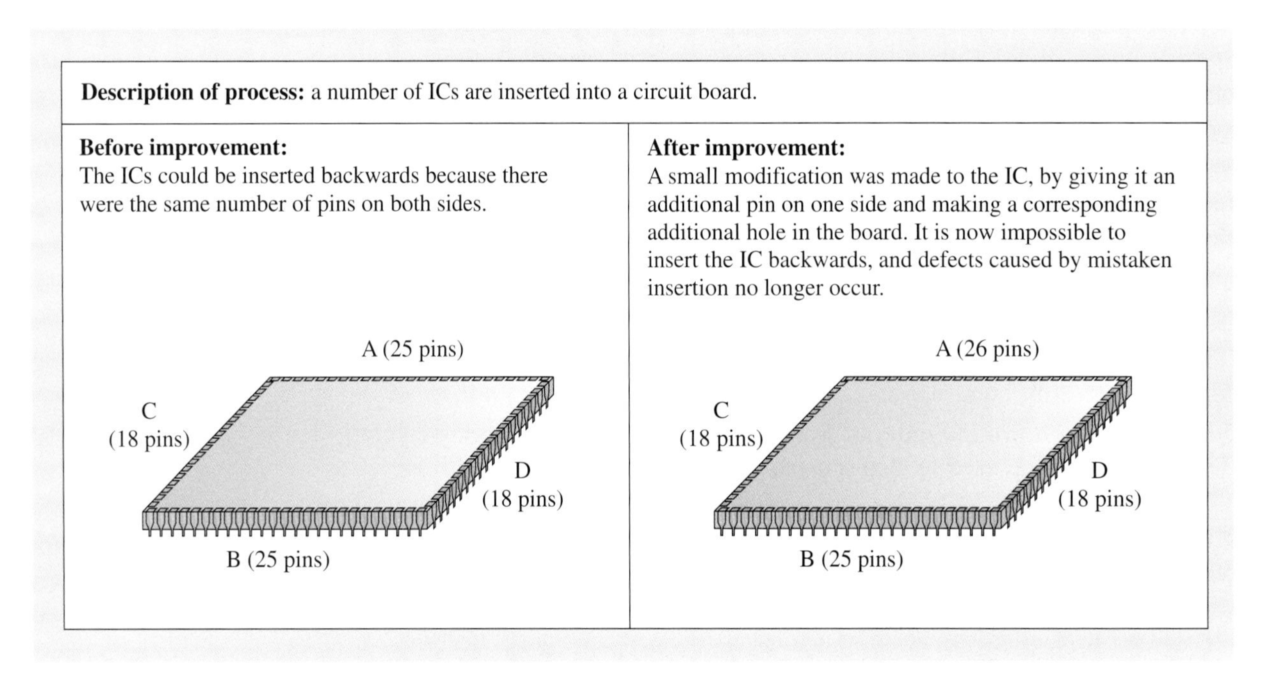

Figure 3.35 A poka-yoke exercise on the mounting of integrated circuits (Source: Nikkan Kogyo Shimbun/Factory Magazine, 1988, p. 174)

3 *By giving a visual or auditory warning*. For example, a temperature sensor and audible alarm were fitted to a cabinet used to sterilise ampoules of water for injection; this enabled operators to be alerted to any dangerous falls in temperature that occurred during the long heating cycle.

Use of *poka-yoke* is often associated with designers and engineers developing new products and processes, but it can also be used as an improvement tool. Although the people concerned might not have recognised the technique by name, or even regarded it as a technique at all, many of the big success stories achieved by shop-floor-level improvement teams have been the result of *poka-yoke*. The very detailed knowledge of processes and possible problems that these teams possess puts them in an excellent position to use the technique, and the results they achieve can make their own jobs easier, as well as eliminating defects.

10.2 Quality function deployment

Quality function deployment was developed as a means of ensuring that customer requirements are translated into appropriate company requirements at each stage of the development of a new product, from research and design through to manufacture and delivery. The technique was developed initially at Mitsubishi's Kobe shipyard in 1972, and it is claimed that use of QFD enabled Toyota, which began using the technique in 1977, to reduce the length of its design cycles by a third and dramatically reduce start-up costs while also improving the quality of its output.

The purpose of QFD is to provide a framework for a team-based approach to the design and development process. The team members will come from marketing, design, engineering, production, quality, purchasing, financial managers, and so on, and they will take a design project forward through four phases using a series of charts. The phases are:

1 product planning
2 parts deployment
3 process planning
4 production planning.

QFD is used in just the same way in problem solving and improvement but the emphasis here is on redesign rather than design.

The house of quality

At the core of the technique is matrix data analysis using a matrix called the house of quality. This matrix is, however, much more than a chart to be filled in; it is a conceptual framework that can be used to guide teams through the transformation process that converts customer requirements into successful products.

A typical form of the house of quality is shown in Figure 3.36. I shall explain the key features of the chart more fully using the numbers given to the various 'rooms' in the figure.

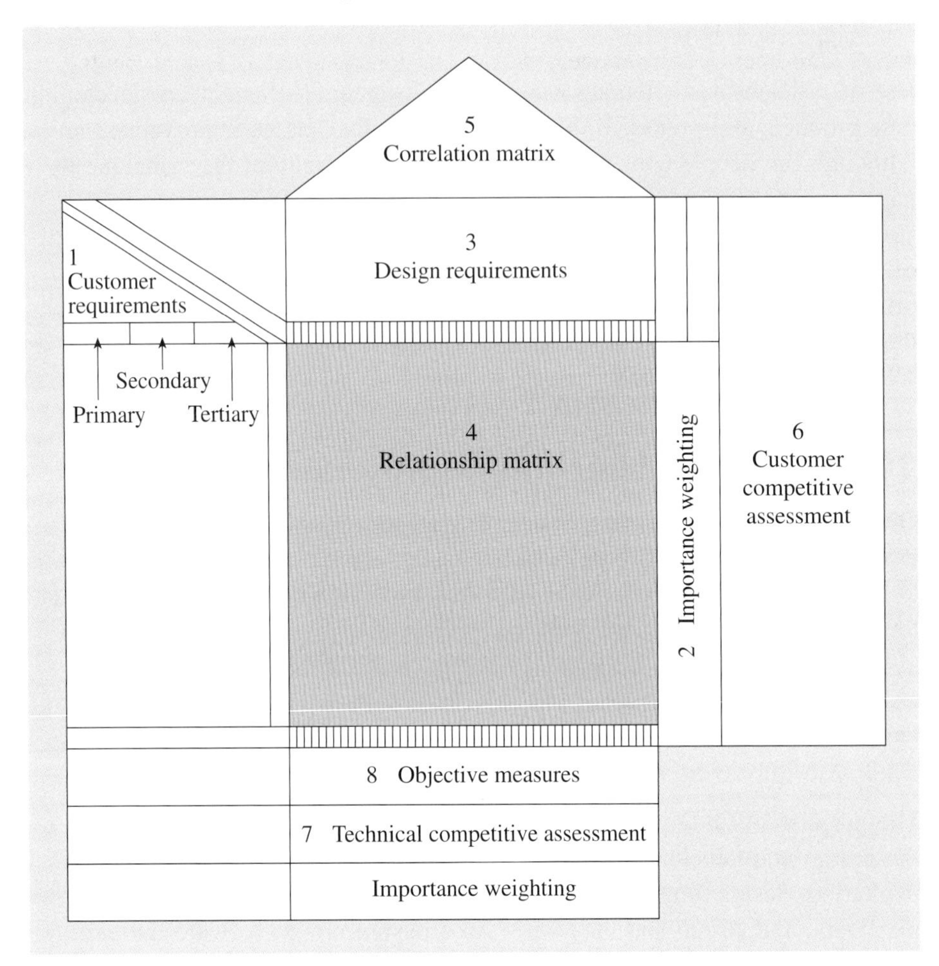

Figure 3.36 The house of quality

1 Customer requirements

The first step of a team using QFD is to work together with colleagues, and with customers and potential customers, to determine customer requirements in a broad sense. Unspoken as well as spoken requirements have to be identified and similar demands consolidated in order to give a list that is complete but of a manageable size. The requirements are usually categorised as primary, secondary and tertiary but the distinctions between these three categories vary between practitioners. For example, some view them hierarchically, so for a car a primary requirement might be 'easy to drive', a secondary 'good all-round visibility', and a tertiary 'good view through rear window'. Others use them as labels for three exclusive sets, such as ease of operation, performance and ease of servicing, or, as you will see in Figure 3.40, quality and suitability of container, and characteristics of content.

2 Importance weighting

Having identified the customer requirements, which typically number between 30 and 100, the next task is to prioritise them in order to indicate the importance of each requirement to the customer. One method is to give each requirement a percentage score so that the total for all requirements adds up to 100. A simpler method is to score the requirements using a five-point scale where 1 = of little importance and 5 = very important.

3 Design requirements

The next stage is to translate the customer requirements (usually expressed as action statements such as 'easy to use'), which are subjective and difficult to transform into action, into design requirements (usually expressed as nouns). These take the form of measurable and controllable specific technical characteristics, which are often referred to as substitute quality characteristics or SQCs. An example of this translation process is shown in Figure 3.37. A customer requirement is translated into SQCs at two levels: the total system expectation level, which is concerned with the final product's performance and which uses an objective measure that should correlate with the customer's perception; and the measurable system parameter level, which identifies variables that directly affect the product's performance but are of no immediate interest to the final customer.

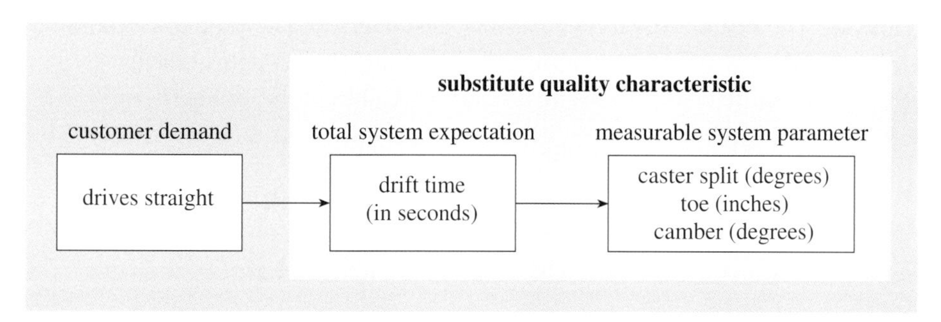

Figure 3.37 The translation of a customer requirement into SQCs (Source: Slabey, 1991)

4 Relationship matrix

The purpose of the relationship matrix is to indicate the direct cause-and-effect relationships between the customer requirements and the design requirements. The strength of the relationship is denoted by a symbol: ⊙ = strong relationship; ○ = moderate relationship; △ = weak relationship. For example, if seat position is identified as a design requirement, it will have a moderate relationship with the customer requirement for a good view through the front windscreen, a strong relationship with the need to be able to use the rear-view mirror easily, and no relationship with a customer requirement for good fuel economy.

Use of the symbols makes it easier to distinguish between the important design requirements (those with many strong relationships) and the less important ones (those with only a few relationships, particularly if those relationships that have been identified are weak).

5 Correlation matrix

The correlation matrix is used to indicate which design requirements support each other (positive correlation) and which are in conflict (negative correlation). Again, symbols are used to denote the nature of the relationship: ⊙ = strong positive; ○ = positive; ╳ = negative; ⋇ = strong negative. On the face of it, positive correlations may sound the most promising because they often indicate the opportunity to make savings by removing implicit duplication. However, negative correlations do in fact sometimes signal the greatest opportunities to gain a competitive edge. Almost all well-designed and well-engineered products require some trade-offs between conflicting requirements, but tremendously successful products often owe their success to innovations that were discovered as a result of trying to reconcile conflicting objectives.

6 and 7 Competitive assessments or benchmarking

Two forms of competitive assessment are used to evaluate the products of various other manufacturers with which the product under discussion would compete. The results of surveys that ask customers to rate products according to how well they meet their needs are placed in 6. A set of technical benchmarks is built up in 7 by obtaining the competitors' products and testing them.

8 Objective measures

Objective measures are the design targets that stem from the construction of the first house. They form the starting point for the next part of the team's work.

Constructing further houses

QFD is a multi-stage technique that progresses through the four phases identified earlier. The output of the first phase is a set of objective measures that relate to the substitute quality characteristics, as you have just seen. This becomes the input to the next phase, parts deployment, in which the design requirements are translated into specific parts including raw materials and components, and the critical characteristics of those parts which would enable the essential functions to be performed are determined.

The part characteristics provide the input to the third phase: process planning. Within the constraints – financial, technological and so on – that are imposed on the design project, this phase seeks to specify the operations that will be needed to manufacture the product. Particular attention is paid to the operations that are most critical in achieving the significant part characteristics and to the process parameters that will have most influence on product quality. This information is then fed into the fourth and final stage, production planning, where the operating procedures are specified.

The inputs and outputs to all four phases are summarised in Figure 3.38. In practical terms, the way in which the output of one stage feeds in as the input to the next is shown as a row of houses in Figure 3.39.

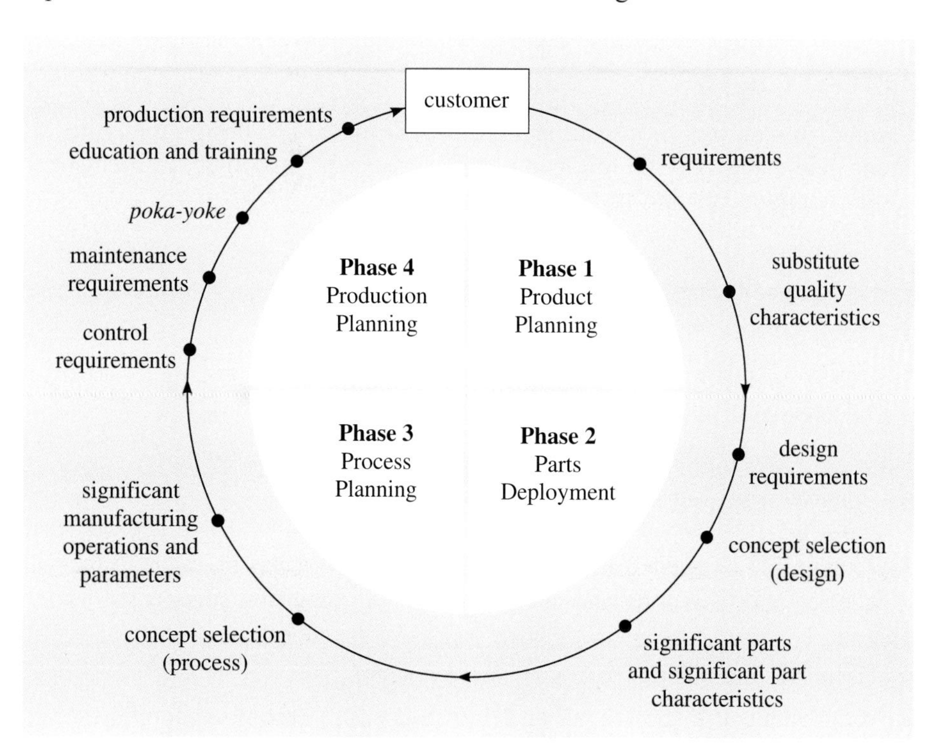

Figure 3.38 The four phases of QFD

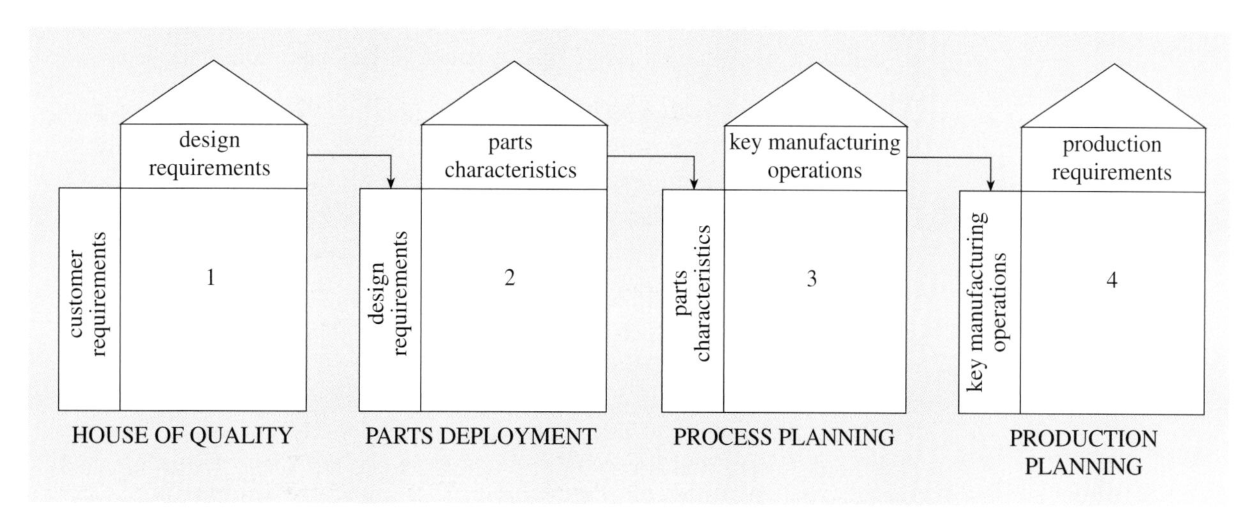

Figure 3.39 The row of houses

An example of a completed house for phase 1 is shown in Figure 3.40, while Figure 3.41 shows examples for all phases. As you can see, the precise layout of the chart and the words used in some of the headings differ from those in Figure 3.36. Different organisations do, in effect, custom-build their houses. Some incorporate information on past customer complaints about similar products, for example; others indicate the degree of technical difficulty associated with meeting the various objective values; and so on.

ACTIVITY 3.16 ....

Assume you are part of a team using the QFD method to design a mobile phone, and categorise the following characteristics as customer requirements, design requirements and parts characteristics:

- lightweight
- low-power microphone
- small batteries
- ability to operate for long periods without recharging
- low power consumption
- CMOS (complementary metal oxide semiconductor) components
- minimal power consumption while not in use. ●

Although QFD originated in manufacturing industry it has spread into the service sector and as long ago as 1995 Terry was reporting success stories such as:

- complete refurbishment of a Japanese hotel
- refurbishment of the second and third floors of a Japanese shopping centre
- improvement of the consumer credit service of the Italian bank Banca Antoniana
- redesign of Alitalia's B747 Intercontinental Business Class service.

Wherever it is applied, a technique such as QFD is, of course, only as good as the people who use it. It offers no miracle solutions, but does provide a framework, a discipline even, to assist people with different areas of expertise to work together to maximum advantage. Further, it facilitates the emergence of the craft knowledge that normally remains hidden because the 'agenda' does not draw it out.

10.3 TRIZ

TRIZ is a technique that has attracted a lot of publicity in recent times but it is difficult to find real-life examples where its use has really led to finished products that are being used by customers or processes that are generating commercial products and services. Nevertheless, I am including it here for two reasons. The first is that, at its core, it has two concepts that are very

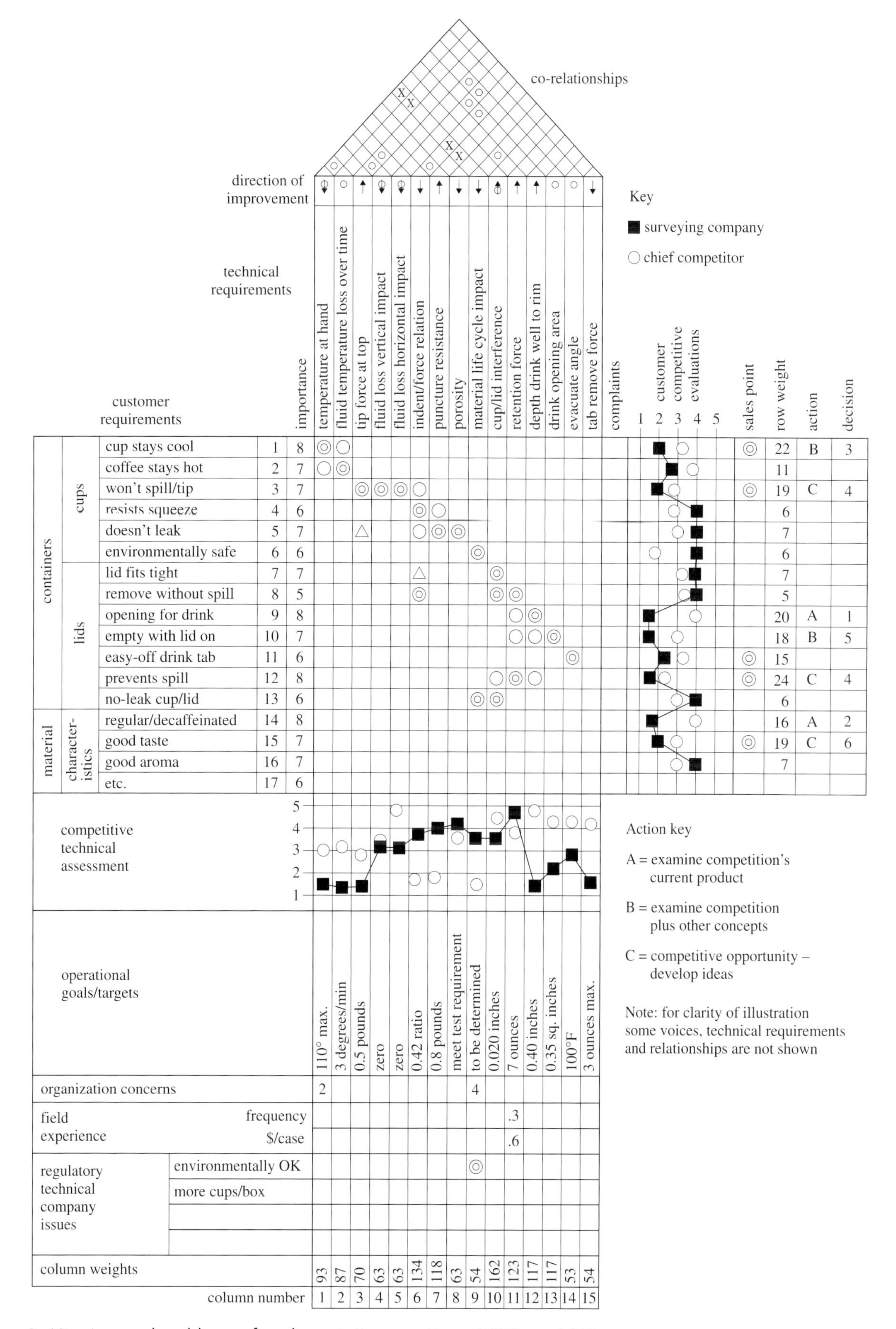

Figure 3.40 A completed house for phase 1 (Source: Day, 1993, p. 105)

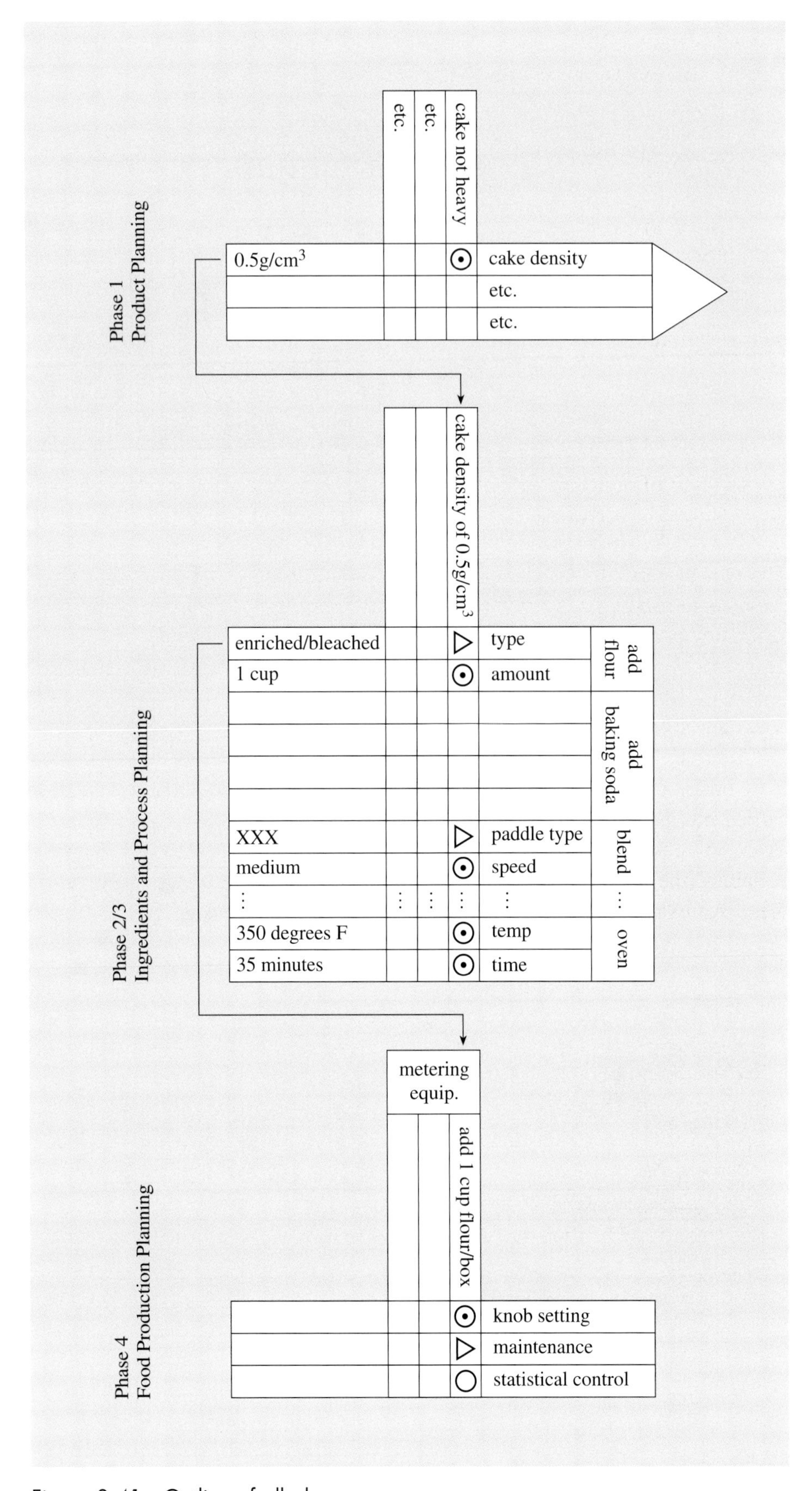

Figure 3.41 Outline of all phases

insightful and valuable. The second is that it is something new that might just take off and produce outstanding results!

Read Offprint 5, which sets out the key aspects of TRIZ.

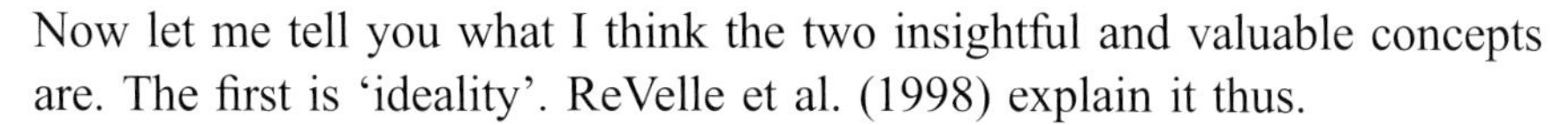

Now let me tell you what I think the two insightful and valuable concepts are. The first is 'ideality'. ReVelle et al. (1998) explain it thus.

> Underlying the unified set theory of TRIZ is the inexorable movement of technical systems towards the Ideal System. The Ideal System, in theory, is a state where the mechanism is absent but the function of the mechanism is present. The Ideal System therefore provides useful function only.
>
> (ReVelle et al., 1998, p. 125)

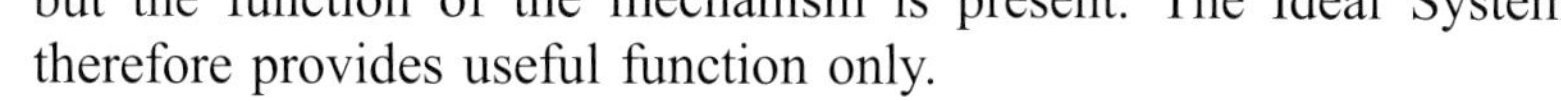

ACTIVITY 3.17 ....

Try to think of an example where a mechanism is absent but the function of the mechanism is present. ●

The second concept is contradictions. According to proponents of TRIZ it provides techniques for eliminating contradictions through the use of separation principles such as separation in time, separation in space, separation in scale and separation between the component and the system.

10.4 Y2X

Y2X builds on the input (I) and output (O) of SIPOC charts, and can be carried out as a precursor to a variety of other techniques designed to bring about process improvements. It uses a matrix diagram to present key output variables (the Ys) and show the key process inputs (the Xs) that influence them. An example of part of an analysis is shown in Figure 3.42.

	X measures			
Y measures	time of day	size of order	new product promotion	number of staff available
queuing time	⊙	△	△	⊙
time to complete order	△	○	○	⊙
availability of seating	⊙		△	○

strong association ⊙ association ○ weak assocation △

Figure 3.42 Y2X matrix – queues at a fast-food restaurant (Source: redrawn from Marconi, 2000, p. 101)

10.5 Taguchi methods

Taguchi methods appeared in the West in the 1980s. In large part their focus was on the optimisation of product and process and on testing design before manufacture. However, they were also recommended as 'a trouble-shooting methodology to sort out pressing manufacturing problems' (Department of Trade and Industry (DTI), 1989, p. 21).

I shall concentrate on just one aspect of Taguchi's work: parameter design experiments. However, because the design, planning and analysis of parameter design experiments require a considerable level of statistical expertise, I aim to do no more than give you a basic understanding of the nature of the technique and the purpose that it fulfils. If you are keen to develop your knowledge further, there are a number of books that you might consult, starting with Taguchi's own work, *Introduction to Quality Engineering* (1986). However, I should warn you that this requires a vastly deeper grounding in statistical methods than that required to study this course. In addition, the statistical software supplied with this course can assist you with the design and analysis of Taguchi experiments.

The central theme of parameter design is to reduce costs by reducing variation. In everyday language, parameter design experiments are trials that are run to discover the optimum specification for a product or the optimum settings for a process, bearing in mind the factors, such as imperfections in raw materials or fluctuations in operating conditions, that tend to cause variations in the output of a process. In more technical language using Taguchi's own terminology: for any product or process a set of performance characteristics can be identified, each with an ideal state called a target value. Performance characteristics are the characteristics of the product or process that determine its ability to satisfy its users' needs. Examples might be the sharpness of the picture on a television set or the output voltage of a circuit. The variables that affect each performance characteristic are divided by Taguchi into two categories: control factors, whose nominal settings define a product or process design specification, and sources of noise, which cause the performance characteristic to deviate from its target value. Products and processes that are insensitive to noise are said to be robust.

Parameter design experiments are conducted in order to identify:

- the settings at which the effects of noise parameters are minimised
- the settings that reduce costs without reducing the level of quality
- the parameters that have a significant effect on the mean values of the performance characteristics but no effect on their variation, so they can be used to adjust the mean values
- the parameters that have no discernible effect on the performance characteristics, so their tolerances can be relaxed.

These objectives are achieved by *systematically* varying the settings of the control factors during an experiment and comparing the effects of noise factors for each test run.

I have italicised *systematically* in the last sentence quite deliberately. Systematically varying the settings of the control factors is one of the key features that lie at the heart of Taguchi methods. Incidentally, this provides an excellent example of Japan's ability to take ideas from the West, use them to great commercial advantage, and then return them with a 'made in Japan' tag attached. Experimental methods of this type were first devised in the 1920s by R. A. Fisher, a statistician who was working on agricultural research at the Rothamsted Experimental Station in Harpenden. Fisher developed some basic principles for the design of experiments and perfected analysis of variance techniques by which it was possible to determine whether the outcomes of different trials were significantly different and to identify those factors to which the differences were attributable. Fisher's methods were, however, deemed to be somewhat cumbersome for industrial use and in the early 1950s Taguchi began to look for alternatives. He found part of his answer in the work on orthogonal arrays that had been published just after the Second World War by Plackett and Burman (1946) and Finney (1945).

The classical method of studying the effects of a number of factors on an outcome is wholly reductionist. The factors are reviewed singly by varying them one at a time while holding the remaining factors constant. If it is thought that the interactions between factors might be important, a full factorial experiment is performed with all possible combinations of factor levels each forming the basis of a separate trial. Thus seven factors, each with two levels, would require $2^7 = 128$ trials and four factors with three levels would require $3^4 = 81$ trials.

ACTIVITY 3.18 ....

How many trials (in round figures) would a full factorial experiment require for 20 three-level factors? ●

Use of what is known as an orthogonal array allows many factors to be studied simultaneously, with the values of many factors (rather than just one) being changed between successive trials. This means that the number of trials needed can be significantly smaller. An example of an orthogonal array is shown in Table 3.10. The ones, twos and threes in the array denote the first, second and third levels of a factor, respectively. Orthogonal arrays have one special property: in every pair of orthogonal columns all combinations of different variable levels occur, and they occur the same number of times. Use Table 3.10 to check this for yourself. This paired balancing property makes a comparison of different test settings of a control factor valid over the ranges

implied by the test settings of the other control factors.
Table 3.11 summarises some of the more commonly used orthogonal arrays.

Table 3.10 The L_9 orthogonal array

Experiment number	Column number and factor			
	A	B	C	D
1	1	1	1	1
2	1	2	2	2
3	1	3	3	3
4	2	1	2	3
5	2	2	3	1
6	2	3	1	2
7	3	1	3	2
8	3	2	1	3
9	3	3	2	1

Table 3.11 Common orthogonal arrays

Orthogonal array	Number of factors	Number of levels per factor	Number of trials required by orthogonal array	Number of trials in a traditional full factorial experiment
$L_4(2^3)$	3	2	4	8
$L_8(2^7)$	7	2	8	128
$L_9(3^4)$	4	3	9	81
$L_{12}(2^{11})$	11	2	12	2 048
$L_{16}(2^{15})$	15	2	16	32 768
$L_{16}(4^5)$	5	4	16	1 024
$L_{18}(2^1 \times 3^7)$	1	2	18	4 374
	7	3		
$L_{36}(2^3 \times 3^{13})$	3	2	36	12 745 584
	13	3		

Another key feature of Taguchi's work is the use of signal-to-noise ratios (*S/N*s) to judge the outcomes of an experiment by identifying the factors that affect variation. In its elemental form, the *S/N* is simply the ratio of the mean to the standard deviation.

Taguchi has developed over 70 different *S/N*s, many of which are unique to specific industries or processes, but for most applications Bendell et al. (1989) recommend that a choice is made from the following three types:

1 Type N for use when nominal is best:

$$S/N_{\mathrm{N}} = 10\log_{10}\frac{\frac{1}{n}(S_m - V_{\mathrm{e}})}{V_{\mathrm{e}}}$$

where S_m is the sum of squares due to the mean $= \dfrac{(\Sigma y_i)^2}{n}$

V_{e} is the variance $= \left(\dfrac{\Sigma y_i^2 - (\Sigma y_i)^2/n}{n-1}\right)$

y_i is an observation

n is the number of observations.

2 Type S for use when smallest is best:

$$S/N_{\mathrm{S}} = -10\log_{10}\left(\frac{1}{n}\Sigma y_i^2\right)$$

3 Type B for use when biggest is best:

$$S/N_{\mathrm{B}} = -10\log_{10}\left(\frac{1}{n}\Sigma\frac{1}{y_i^2}\right)$$

It is worth noting that Taguchi has been criticised for his use of *S/N* as a performance statistic. His critics argue that it is preferable to study the mean and the variance separately rather than combining them in a single function.

I shall now outline the stages of a typical parameter design experiment. In doing this, I shall draw on one particular application of the technique as a source for my examples. The application concerns an experiment carried out on the placement of surface mount components on circuit boards at Mars Electronics, which was reported in a paper by Bandurek et al. (1988).

Outline of the parameter design experimental method

1 Select the area for experimentation

It is fair to say that many companies have failed in their attempts to use Taguchi methods because they have made poor decisions about where to start their programmes. They have looked for a problem on which to try Taguchi methods rather than starting with a problem and then considering whether a parameter design experiment might provide an appropriate means of finding a solution. Given this general warning, there is also one very important specific proviso when making the selection: you cannot conduct a Taguchi experiment on a product that you cannot make to the required standards or on a process that you cannot control.

Now I shall introduce you to the problem that Mars Electronics selected. The company uses surface mount technology in its production of coin validation equipment and small radars. Surface mount components offer a significant advantage over their more traditional counterparts because their smaller dimensions allow much higher component densities on printed circuit boards (PCBs). (The leg spacing of a surface mount integrated circuit (IC) is normally 1.27 mm compared with a spacing of 2.54 mm for a conventional IC.) The use of surface mount components does, however, have a corresponding drawback: they must be positioned on the PCBs with much greater precision. A pre-experiment investigation at Mars Electronics revealed that variability in the performance of a numerically controlled placement machine was giving cause for concern, so the company decided to focus on this problem.

2 List the performance characteristics, the control factors and the sources of noise

Most products and processes have a number of performance characteristics, so it is important to identify those that are pertinent to the experiment and to devise some measure(s) of performance.

At Mars Electronics, Bandurek et al. were concerned to investigate accuracy of placement and thus had to find a way of measuring this. They selected eight distances (shown as A–H in Figure 3.43) and calculated the differences between the pairs A and D, E and H, B and G, and C and F to give the displacements of the edges of each IC. The overall inaccuracy of positioning for each IC was then represented by the worst of the four displacements.

ACTIVITY 3.19 ....

In their experiment Bandurek et al. wished to minimise the inaccuracy of IC positioning. Of the three S/Ns that were listed earlier, which would you expect them to choose? ●

Because most products and processes have a large number of control and noise factors, usually they cannot all be studied in a single experiment. An additional restriction is the number of levels of each that are used; these are usually limited to two or three. Idea generation techniques (which I describe in Section 8) are used to identify the factors that are likely to affect the product or process and to classify them as either control or noise factors.

The factors that were finally chosen after much discussion in the Mars Electronics application are shown in Table 3.12. The numbers in brackets indicate the number of levels used for each factor. A further example drawn from the automotive industry is shown in Table 3.13; the aim of this study was to reduce the assembly effort of a fuel systems spring lock connector while maintaining sufficient pull-off force requirements. Don’t worry about

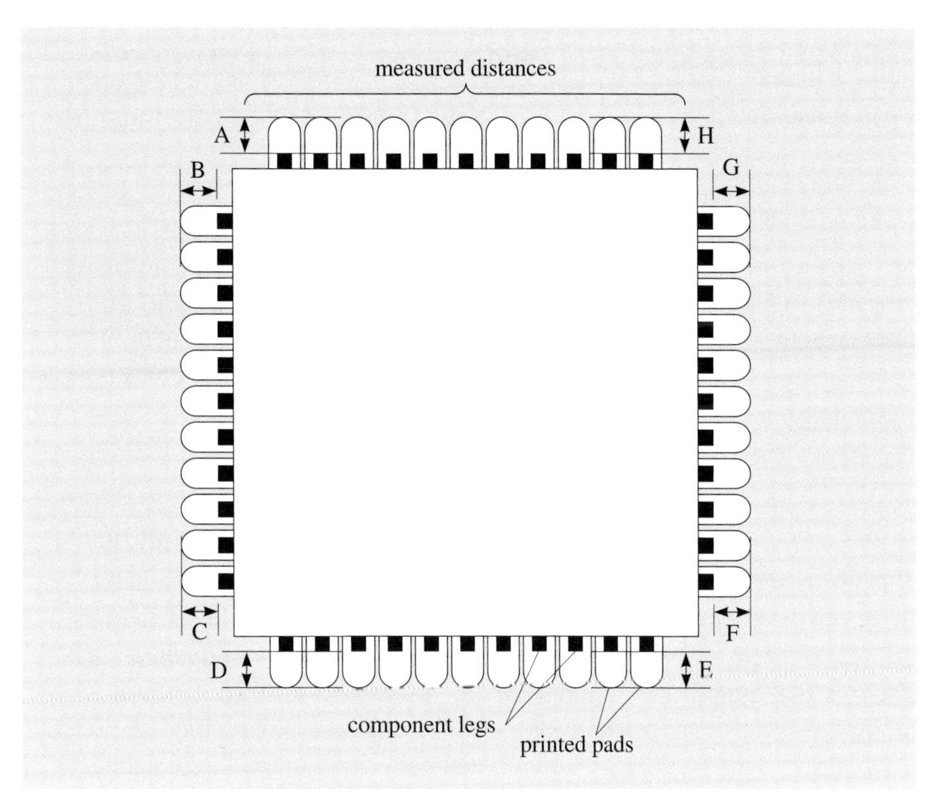

Figure 3.43 Placement accuracy (Source: Bandurek et al., 1988, p. 173)

the detail of this table. I have included it merely to give you a further illustration of the sorts of factor that it might be appropriate to include.

Table 3.12 Control and noise factors – Mars Electronics

Control factors		**Noise factors**	
Speed of convergence of jaws	(2)	Board on panel	(2)
Pressure of squeezing jaws A	(3)	Component type	(2)
Pressure of squeezing jaws B	(3)	Track – component type 1	(3)
Placement force	(3)	Track – component type	(3)
Tweezer mode	(2)	Time trend	(4)

(Source: Bandurek et al., 1988, Table II)

Table 3.13 Control and noise factors – spring lock connector experiment

Control factors		**Noise factors**	
Indicator ring	(2)	Amount of lubricant	(2)
Lubrication type	(2)	Concentricity	(2)
70°–100° ∢ – FSLC	(3)	Runout	(2)
Spring wire diameter	(3)	'O' ring volume	(2)
Flare diameter – FSLC	(3)		
Gland width – MSLC	(3)		
FSLC ID lubrication	(3)		
Cage/spring ramp configuration	(3)		

(Source: adapted from Ciucci, 1988)

3 Construct the design and noise matrices and plan the parameter design experiment

This is the stage at which considerable expertise in the use of experimental statistics is required. Two orthogonal arrays must be selected: an *inner array* of control factors and an *outer array* of noise factors. Figure 3.44 shows two small arrays that might be suitable in a simple experiment. The columns of the inner array represent the control factors and its rows represent the different combinations of test settings. The columns of the outer array represent the noise factors and its rows represent different combinations of noise levels.

Planning the experiment is a two-part activity: it is necessary, first, to decide how to vary the control factors and, second, to decide how to measure the effects of noise. The complete experiment requires the two matrices to be combined so that each test run of the control factor matrix is crossed with all rows of the noise factor matrix. Thus the number of runs necessary if using the two arrays in Figure 3.44 is 36 (that is, 9×4).

Tables are available to aid the selection of arrays (see Wu, 1986, for example), but even so the choice is not always straightforward. For example, the inner array must be large enough to include all the levels of the control factors with the most levels, but 'dummy' levels will have to be used to replace some of the levels for those factors with fewer levels. Another major problem concerns the randomisation of the order of the trials. Randomisation is recommended in order to minimise the effects of unknown variables, but the nature of the production process may be such that it is very difficult to jump between various settings at random, for example when one of the factors is oven temperature.

The arrays used by Bandurek et al. are shown in Figure 3.45; you don't need to concern yourself with the detail of this figure.

outer array – noise factors

3	2	1	row no.
1	1	1	
1	2	2	
2	1	2	
2	2	1	

inner array – control factors

row no.				
1	1	1	1	1
2	1	2	2	2
3	1	3	3	3
4	2	1	2	3
5	2	2	3	1
6	2	3	1	2
7	3	1	3	2
8	3	2	1	3
9	3	3	2	1

Figure 3.44 Two small arrays that may be used in a parameter design experiment

The following is a quote from Bandurek et al.'s paper to indicate how they tackled their problems of array selection and planning:

> After consulting the table of common orthogonal arrays ..., the $L_{16}(4^5)$ orthogonal array was selected for both the inner array of control factors and the outer array of noise factors; this being the smallest orthogonal array capable of coping with this number of factors and levels without combining factors. The total number of trials required in the experiment was therefore $16 \times 16 = 256$ trials, each combination of control factors being run at each combination of noise factors.
>
> The major disadvantage of the $L_{16}(4^5)$ design was the number of 'dummy' levels required for this experiment. Where the experiment required only three levels for a factor, the fourth level in the orthogonal array was set equal to the central (nominal) level. Where only two levels were required, each factor level was itself assigned two levels in the orthogonal array. ...

outer L_{16} array – noise factors

16	15	14	13	12	11	10	9	8	7	6	5	4	3	2	1	row no.
4	4	4	4	3	3	3	3	2	2	2	2	1	1	1	1	F
4	3	2	1	4	3	2	1	4	3	2	1	4	3	2	1	G
1	2	3	4	2	1	4	3	3	4	1	2	4	3	2	1	H
3	4	1	2	1	2	3	4	2	1	4	3	4	3	2	1	I
2	1	4	3	3	4	1	2	1	2	3	4	4	3	2	1	J
2	2	2	2	1	1	1	1	2	2	2	2	1	1	1	1	F
A	M	A	M	A	M	A	M	A	M	A	M	A	M	A	M	G
A	B	C	A	B	A	A	C	C	A	A	B	A	C	B	A	H
C	A	A	B	A	B	C	A	B	A	A	C	A	C	B	A	I
2	1	4	3	3	4	1	2	1	2	3	4	4	3	2	1	J

inner L_{16} array – control factors

row no.	A	B	C	D	E	A	B	C	D	E	16	15	14	13	12	11	10	9	8	7	6	5	4	3	2	1	mean	S/N ratio
1	1	1	1	1	1	C	30	30	30	NW	21	11	13	3	6	8	14	13	20	9	16	8	8	10	9	8	11.06	–21.60
2	1	2	2	2	2	C	35	35	40	BT																	8.75	–20.18
3	1	3	3	3	3	C	45	45	45	NW																	9.31	–20.21
4	1	4	4	4	4	C	35	35	40	BT																	9.19	–20.19
5	2	1	2	3	4	S	30	35	45	BT																	9.00	–20.68
6	2	2	1	4	3	S	35	30	40	NW																	9.34	–20.76
7	2	3	4	1	2	S	45	35	30	BT																	8.34	–19.17
8	2	4	3	2	1	S	35	45	40	NW																	9.63	–21.07
9	3	1	3	4	2	C	30	45	40	BT																	11.50	–22.23
10	3	2	4	3	1	C	35	35	45	NW																	9.44	–20.57
11	3	3	1	2	4	C	45	30	40	BT																	10.25	–21.35
12	3	4	2	1	3	C	35	35	30	NW																	11.50	–22.09
13	4	1	4	2	3	S	30	35	40	NW																	8.75	–20.75
14	4	2	3	1	4	S	35	45	30	BT																	8.88	–20.79
15	4	3	2	4	1	S	45	35	40	NW																	8.81	–20.28
16	4	4	1	3	2	S	35	30	45	BT																	8.44	–19.79

Figure 3.45 The arrays used by Bandurek et al. (Source: Source: Bandurek et al., 1988, p. 176)

An alternative approach would have been to use five columns of the L_{18} orthogonal array for the control factors, requiring only one 'dummy' level, but this would have increased an already relatively large experiment to $18 \times 16 = 288$ trials.

... The ideal method of conducting the experiment would have been a complete randomization of all 256 trials, but the time trend noise factor prohibited this if the experiment was to be completed in a reasonable time. For operational reasons, it was decided to conduct the experiment over two days with 32 trials in each time trend period per day. As the speed of convergence of the jaws was very difficult to reset and the tweezer mode difficult to change it was decided to block the experiment over these variables.

(Bandurek et al., 1988, pp. 174–5)

4 Run the experiment and use the results to predict improved control parameter settings

As a result of running their experiment, Bandurek et al. drew three conclusions about the way in which parameter design experiments should be conducted:

[A]ll equipment and machinery to be used should be fully overhauled prior to the experiment, time should be allowed for the experiment to over-run, and statistical expertise should be available whilst conducting the experiment so that if changes had to be effected, they would have minimal effect on the randomisation.

(Bandurek et al., 1988, p. 175)

The first step in the analysis of the experimental data is to calculate the mean response for each row of the inner array. This is done using the formula:

$$\bar{y}_i = \frac{1}{n} \sum_{j=1}^{n} y_{ij}$$

where y_{ij} denotes the response for row i of the inner array and row j of the outer array. The next step is to calculate the appropriate *S/N* ratios. Both sets of results can then be used to assess the effects of the various control factors. Figure 3.46 shows the results obtained at Mars Electronics.

Taguchi (1986) suggests that 'the combination [of factors] that gives the better signal-to-noise ratio should be selected, even when the difference is slight', but this piece of general advice does, again, highlight the need for both statistical expertise and a sound, detailed knowledge of the product and/or process in question. For example, Bandurek et al. knew that it was not desirable to run the placement machine at Mars Electronics at extreme pressure values, so rather than take Taguchi's advice at face value they decided to conduct further analysis of their data in order to determine the statistical significance of the effects of the factors on the mean response and the *S/N*s. When this analysis yielded disappointing results, they carried

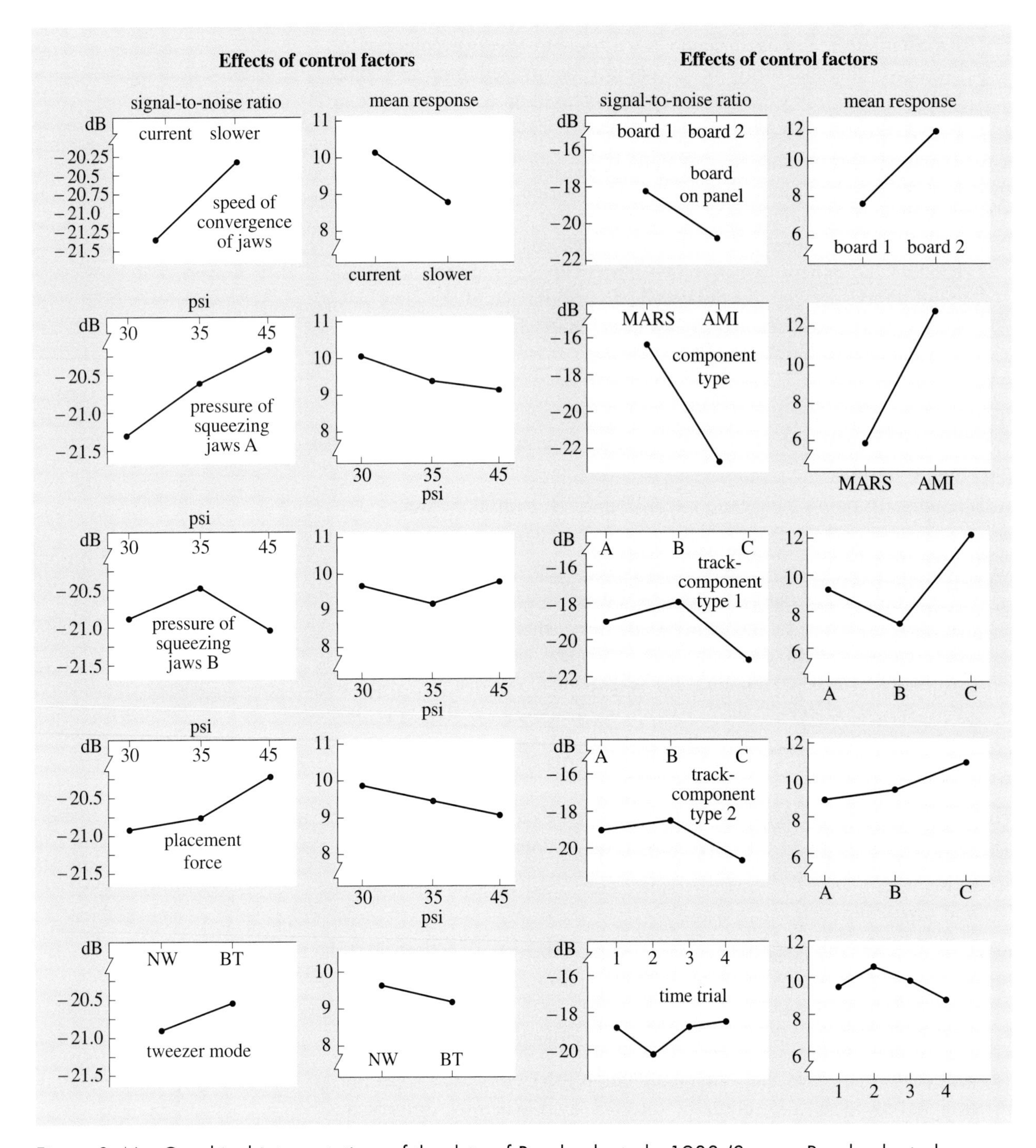

Figure 3.46 Graphical interpretations of the data of Bandurek et al., 1988 (Source: Bandurek et al., 1988, p. 180)

out a similar analysis on the noise factors. Their conclusion was that their experiment had yielded optimum control settings and that it had, incidentally, revealed an error in the programming of the placement arm.

5 Run a confirmation experiment to check the prediction

After running a parameter design experiment, it is necessary to confirm that the settings suggested by it will indeed result in an improvement. As much as anything, the purpose of this stage is to check that the statistical model that underpinned the design and analysis was in fact valid.

Taguchi methods are by no means the only experimental designs that are available. Indeed, some would argue that other options are superior, especially where their efficiency is concerned. The statistical software includes some other experimental design options. I therefore urge you to undertake the following computer activity even though it is an optional extra.

Now do Exercise 3.1 in the Computer Exercise Booklet.

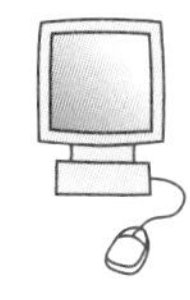

10.6 SMED

SMED is a technique that is closely associated with the lean methods you will be meeting in Block 4, but even used on its own it is capable of delivering improvements and helping to solve throughput problems.

SMED is often associated with batch manufacture where its purpose is to reduce the time that elapses between finishing one batch and reaching full production speed on the next batch. It is this batch manufacturing association that gives SMED its original name: single minute exchange of dies. However, SMED can also be used in many other settings. For example, in the transport industry it can be used to try to minimise the time taken to prepare for the next journey. (It could have been used to support South-West Airlines in its effort to reduce turnaround time for aircraft that was mentioned in Section 9.4.)

The process for using the SMED technique is as follows:

1. Observe the changeover closely, making detailed notes of what is done and how long each activity takes. Depending on the size of the operation it may be possible to prepare a video recording for later analysis; alternatively a team of people may be needed to carry out the observation.
2. Analyse the data by identifying those activities that can be undertaken only when the process is not operating. These are classed as internal activities. Shifting the remaining activities (the external activities) so that they are undertaken while the process in under way is likely to reduce the changeover time by 15 to 20% for a typical process.
3. Find ways of redesigning internal activities so as to reduce the amount of time they take or, better still, convert them into external activities. For example, if a machine needs to cool down before it can be reset, finding a way of cooling it more quickly will reduce the time of the internal activity.

4 Streamline the external activities so as to make efficiency gains. For example, tools required to carry out a particular activity can be moved closer to the place where the activity is undertaken.

5 Document the revised process and put controls in place to monitor future performance.

10.7 Process redesign

The starting point for process redesign is to model the current process. The four diagram types you met in Sections 6.1 to 6.4 are very useful for this: activity sequence diagrams, flow-block diagrams, flow-process diagrams and spaghetti charts. The purposes of these are to:

- identify the different stages of the process
- identify physical flows and flows of information
- provide evidence of excessive amounts of movement and/or poorly designed layouts
- identify any bottlenecks or superfluous stages.

For a service process it is also important to identify the points at which contacts with customers (internal and external) take place. The reason for this is that measures of the customers' perceptions of the service contact process are likely to be qualitative rather than quantitative and to relate to things like friendliness, courtesy, professionalism, extent to which confidence is inspired, and so on. The provision of some of these might be at odds with the desire to give the most efficient service, so it is particularly important to be clear about objectives when looking at these stages of a process. Aside from these considerations the best opportunities for improving processes are likely to centre on maximising capacity and eliminating waste in all its various forms. Block 4 will look at approaches designed to do this but in the meantime I shall consider just two aspects: throughput time analysis and bottleneck analysis.

Throughput time analysis requires tabulation of the processing times at each stage of the process, the times taken to travel between stages, and all the waiting times, so that apparent opportunities for improvement can be identified. A bottleneck is the stage of a process that constrains the process's capacity. Bottlenecks thus determine the rate of output that processes are capable of achieving. Capacity cannot be increased without removing a bottleneck, but of course removing one bottleneck causes the stage that had the second lowest throughput rate to become the new bottleneck.

It is very unlikely that the capacity of all the stages of a complex process can be balanced perfectly, so the treatment of bottlenecks is very important. One type of production-scheduling software package focuses on bottlenecks. It is called optimised production technology (OPT) and it aims to plan production in such a way as to minimise the problems associated with bottlenecks. It is worth looking at its principles because they have wider application than the scheduling of manufacturing production with which they are normally associated. Slack et al. (2007) set out the underlying principles of what they call this 'improvement-oriented approach' thus:

1. Balance flow, not capacity. It is more important to reduce throughput time rather than achieving a notional capacity balance between stages or processes.
2. The level of utilization of a non-bottleneck is determined by some other constraint in the system, not by its own capacity. This applies to stages in a process, processes in an operation and operations in a supply network.
3. Utilization and activation of a resource are not the same. According to the TOC [theory of constraints] a resource is being *utilized* only if it contributes to the entire process or operation creating more output. A process or stage can be *activated* in the sense that it is working, but it may only be creating stock or performing other non-value-added activity.
4. An hour lost (not used) at a bottleneck is an hour lost for ever out of the entire system. The bottleneck limits the output from the entire process or operation, therefore the under-utilization of a bottleneck affects the entire process or operation.
5. An hour saved at a non-bottleneck is a mirage. Non-bottlenecks have spare capacity anyway. Why bother making them even less utilized?
6. Bottlenecks govern both throughput and inventory in the system. If bottlenecks govern flow, then they govern throughput time, which in turn governs inventory.
7. You do not have to transfer batches in the same quantities as you produce them. Flow will probably be improved by dividing large production batches into smaller ones for moving through a process.
8. The size of the process batch should be variable, not fixed. Again, from the EBQ [economic batch quantities] model, the circumstances that control batch size may vary between different products.

9. Fluctuations in connected and sequence-dependent processes add to each other rather than averaging out. So, if two parallel processes or stages are capable of a particular average output rate, in parallel they will never be able to achieve the same average output rate.
10. Schedules should be established by looking at all constraints simultaneously. Because of bottlenecks and constraints within complex systems, it is difficult to work out schedules according to a simple system of rules. Rather, all constraints need to be considered together.

(Slack et al., 2007, p. 457)

11 STATISTICAL PROCESS CONTROL

This section aims to explain statistical process control (SPC), and to teach you how to apply its tools in order to solve problems in processes of all kinds.

The centre of SPC's attention is variation in process performance. Variation may be the problem itself, for instance when a process cannot reliably deliver a desired level of quality. In such cases, control actions suggested by SPC can reduce variation and lead to improved process efficiency and effectiveness. In other cases, variation may be a symptom of another, deeper problem and SPC's role then is to reveal patterns within the variation, providing information about how the process is operating. If a problem is not related to process variation, SPC is not the right tool to solve it.

One of the most important pieces of information revealed by SPC is the extent of two kinds of variation, special-cause variation and common-cause variation. A special cause is something unusual such as a machine breakdown or a condition not previously encountered, typically associated with words like 'fault', 'error' and 'disturbance'. Special-cause variation produces abnormal process outcomes and makes a process unstable. In contrast, a common cause, such as a change in ambient temperature or the rate at which people ask for a service, is inherent in a process, associated with words like 'expected' or 'inevitable'. Common-cause variation therefore produces normal process outcomes. It is a problem only if the amount of variation is greater than people want, in which case it renders the process incapable of meeting their requirements. In general, instability is an operational problem while incapability is a strategic problem.

ACTIVITY 3.20

This activity asks you to watch and consider the content of Programme 1 on the course DVD. It looks at three different contexts in which SPC has been applied and three different reasons for applying it (summarised in Table 3.14). Before watching the programme, think about which of these might be closest to your own situation.

As you watch, pause to make notes on anything that you find personally relevant, and also make general study notes on the following topics:

- uses and potential benefits of SPC
- process variation, including the contrast between common and special causes
- process capability, including its estimation by frequency chart
- control charting.

Table 3.14 Content of Programme 1

Type of process	Context	Purpose
Medical diagnostic procedure	Children's hospital	Understand the process and its limitation
Component manufacturing	Engineering firm	Maintain control over the quality of process output
Service delivery	Police force	Diagnose low process performance

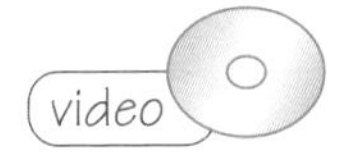

You should now watch Programme 1 *Managing processes: SPC in action*, which is on the course DVD.

SPC has become a lot easier to carry out since the advent of statistical software, such as that supplied with this course. The software has automated SPC's calculations and procedures, which used to be laborious when done by hand.

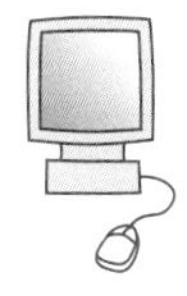

Brief exercises with the software's SPC tools appear throughout this section, so you will need access to the software as you study it.

11.1 Selecting processes for SPC application

One method for selecting processes for SPC application is suggested by Goh et al. (1998). It is based on the concept of process criticality, which has two aspects: technical criticality and statistical criticality. In everyday words, technical criticality assesses the importance of a process in terms of what might go wrong and how serious the consequences would be. Statistical criticality, on the other hand, assesses importance in terms of how badly the process is performing.

Technical criticality

A process is deemed to be technically critical if its failure to conform to specifications would ultimately result in defective products or services. This applies whether or not such a failure has yet occurred. The extent or seriousness of technical criticality depends on the potential impacts of a failure. No standard definition exists of which impacts to include, who is impacted, or how impacts are measured: these are left to evaluators. One way to measure impacts might be to apply the severity classification of FMEA, which was discussed in Section 7.1.

Evaluators may find it useful to categorise the various specifications to which a process should conform, in order to structure priorities. Table 3.15 is loosely based on categories used by Goh et al. (1998).

Table 3.15 Categories of technical criticality

Specification describes	Critical effect	Measured in	Example
Function	Unfitness for purpose, such as malfunction in use	Typically a feature of the product or service	
Level of service	Inadequate availability, reliability, sustainability, etc.	Typically a feature of delivery or usage	
Customer perception	Rejection by customers without functional or service level failure	Perceived appeal, benefits and their opposites	
Downstream effect	Rejection by or harm to non-customers, e.g. machine wear, subsequent process malfunction, pollution	Impacts on subsequent processes, third parties and the external environment	

In Programme 1 the process for answering urgent calls to the police would be deemed technically critical. The presence of a police officer is the functional output of the process, whereas the time to arrive is a service level. Nevertheless, it is a critical service level because it might mean the difference between life and death.

ACTIVITY 3.21

Fill in the example column of Table 3.15, basing your responses on a selection of processes with which you are familiar. ●

Statistical criticality

A process is deemed to be statistically critical when measurements show that it is unstable, or incapable of meeting its performance specifications, or both.

An unstable process is one with abnormal variation in its outcomes. Abnormal variation is revealed by fluctuations, such as step changes, oscillations or spikes, in charts drawn from data about the outcomes, and by statistics calculated from the data that suggest the fluctuations are unlikely to have happened by chance. Abnormal fluctuations usually indicate that special causes of variation are disturbing the process.

An incapable process is one whose variation causes outcomes to fall too often outside specification limits or tolerances. Typically, this means that the process produces too many rejections or defects. Process capability is investigated by comparing the variation *specified* between tolerances with the *actual* variation in outcomes. When actual variation exceeds specified variation, the process is incapable.

Notice that, using these definitions, an unstable process may or may not be incapable, and an incapable process may or may not be unstable.

Estimating statistical criticality

Statistics require data and SPC requires data describing its processes. This requirement may be particularly difficult to satisfy when *estimating* statistical criticality. Logically, if estimation is necessary, the work is likely to be at an early stage at which there is probably no established infrastructure to collect data and hence no data readily available or well organised. For initial estimates, data may have to be collected in a one-off study or gleaned from old records.

Before data are available, it may be possible to make qualitative judgements. A process may have a reputation for stability or a sad history of defects; or, better, it may have no reputation at all because it never gives trouble; or, worse, it may be known for occasional nasty surprises. Qualitative judgements may therefore give an early impression of candidate processes but they are not reliable enough for a systematic assessment of statistical criticality.

Estimating process stability

In SPC, a technique called control charting is used to investigate process stability. The technique is especially good for detecting the fluctuations of an unstable process. The next exercise generates a control chart.

Now do Exercise 3.2 in the Computer Exercises Booklet.

The software would flag in the control chart any abnormal patterns that indicate process instability. You will study such patterns in more detail later.

Estimating process capability

There are two ways to estimate process capability, both involving a comparison between specified variation and actual variation. The comparison can be estimated from a chart, or it can be calculated as a ratio between the two quantities called a process capability index.

Frequency charts can be used to estimate process capability. To compare actual variation with specified variation all you need to do is to assess whether and by how much the columns of the chart lie outside the limits specified for the process in question. Sometimes, however, a frequency chart hinders the comparison. Actual data values are rarely distributed smoothly. The columns can therefore appear very jagged and the edges of the columns may not coincide exactly with the limits, making it difficult to tell how many values in an interval are inside the limit and how many are outside. One way to improve the comparison is to smooth out the frequency chart by drawing its normal probability distribution. The normal curve represents an idealised distribution of values; the mean and standard deviation of the fitted curve are the same as the mean and standard deviation of the sample. As you saw in Block 2, the statistical software can produce this curve for you in a few seconds.

Now do Exercise 3.3 in the Computer Exercises Booklet

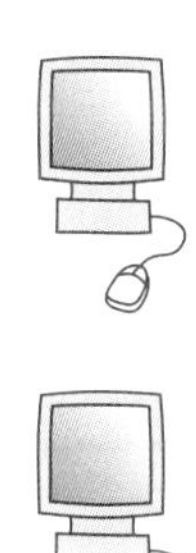

So process capability can be estimated visually from the extent to which a sample of data lies within specification limits or tolerances. The other way to estimate process capability is to calculate it.

Now do Exercise 3.4 in the Computer Exercises Booklet.

You will have noticed several indicators of process capability in the results of this exercise, and may have wondered how to choose between them. With the parts data this seems not to matter, because you would be likely to draw identical conclusions from each of the indicated values. However, this is not always the case.

The most appropriate indicator is governed by the context of the calculation. The goal at the stage I am discussing is an early estimate of process capability, nothing more. At such an early stage you do not know whether the processes being examined are stable, even if control charts indicate they may be. When a process is unstable, the capability indices associated with the between-subgroup and within-subgroup standard deviation (that is, the 'C' ratios) are unreliable. Unless stability has been demonstrated statistically, it is best to estimate process criticality from the capability indices associated with the overall sample standard deviation (that is, the 'P' ratios). I shall look at the use of the C ratios of a stable process later in this section.

In general the ratios are more appropriate for comparing processes than the expected performance. This is because people's interpretation of 'parts per million' performance can be coloured by their knowledge of actual processing volumes, whereas the ratios are simply dimensionless numbers. Expected performance can be helpful, however, when comparing processes whose outcomes have different frequency distributions; in other words, their histograms exhibit significantly different shapes.

The overall capability ratios are:

Pp The specification spread (upper specification limit or USL minus lower specification limit or LSL) divided by the spread between control limits (6 times? overall standard deviation). Pp does not take account of the shift you observed in the histogram in Exercise 3.4. It measures only the width of spread.

PpL The lower within-specification spread (process mean minus LSL) divided by half the spread between control limits (3 times? overall standard deviation). This does take account of the shift in the histogram because it uses the mean value.

PpU The upper within-specification spread (USL minus process mean) divided by half the spread between control limits. PpU and PpL are twin metrics either side of the mean.

Ppk The lesser of PpL and PpU. This ratio represents the worst spread in the data.

Cpm Calculated only if a target value for the process is provided; indicates how closely the process mean matches the target. The target must be between the specification limits.

ACTIVITY 3.22

If you were looking at a range of processes and deciding where to focus your improvement efforts, which of the above capability ratios would you use in order to compare the performance of the processes? ●

SPC practitioners use various rules of thumb to interpret performance and capability ratios. A ratio of 1.33 or higher is deemed acceptable. Given the way the capability ratios are calculated, 1.33 means that at least four standard deviations of the data lie within acceptable limits. Anything less really ought to be improved. A ratio of 2 indicates the well-known Six Sigma (six standard deviations) capability which gave birth to an approach you will study in Block 4.

Thus the parts data indicates a thoroughly incapable process. First, this is due to the shift of the data towards the upper specification limit, which produces a tiny PpU and hence a tiny Ppk. Second, even if this shift were corrected, Pp would still be too small, suggesting that further action would be needed to reduce process variation. In visual terms, the histogram needs to move to the left and to become slimmer, until the tails of the normal curve only just intersect with the specification limits. Such a shape is the hallmark of a process capable of meeting its specifications.

Prioritising processes

The first processes to deal with are any that are not specified well enough to assess their criticality. They are not ready. Before applying SPC to them, it is necessary to specify what conformity to specification means in each case, decide how to measure it, and collect measurements.

Once process criticality has been estimated, prioritisation is a relatively simple matter. Goh et al. (1998) suggest combining the two dimensions, as in Figure 3.47.

Group 1 processes are both statistically critical – there is too much variation now – and technically critical – the consequences of variation could be damaging. They are prime candidates for an initial implementation of SPC. The aim would be to diagnose the causes of variation, to eliminate these as far as possible, and to maintain control over special causes thereafter. You will study methods for doing so in Section 11.3.

		statistically	
		critical	non-critical
technically	critical	**Group 1** need immediate attention	**Group 2** need later attention
	non-critical	**Group 3** need later attention	**Group 4** good

Figure 3.47 Classifying processes into four groups (Source: adapted from Goh et al., 1998, p. 67)

Processes in groups 2 and 3 are critical in only one dimension, and therefore have less priority. Nevertheless, those in group 2 are deemed to influence final product quality significantly, while those in group 3 produce excessive variation even though it is not damaging. Therefore, these groups should eventually be managed by means of SPC.

Group 4 processes are not critical in any way. They can be omitted from an initial implementation, but it is useful to record the rationale for excluding them in case the organisation needs to look at them in the future.

On a large scale, where there are many processes in a group, a further level of prioritisation may be called for. This is unlikely to be an easy decision, because the processes being compared will normally have different metrics of conformity. General approaches include the following:

1 Prioritise small values of Ppk. Estimated process performance ratios are among the few metrics likely to be readily available at this stage of SPC implementation and common to all processes. They implicitly place statistical criticality above technical criticality. A drawback, however, is that effort may be directed to some processes that offer relatively little payoff.

2 Prioritise the highest rate (or risk) of loss due to non-conformity. This places technical criticality above statistical criticality. A drawback is that effort may be directed to some processes that suffer from relatively little variation.

3 Use a multi-variable prioritisation method, such as the 'analytic hierarchy process' outlined in Goh et al. (1998).

4 Assign the highest priority to the processes that operate earliest in the organisation's chain of activities (the value chain) and the operational sequences within it. The rationale is that, if you reduce variation in the outcome of early processes, you may improve the behaviour of later dependent processes, which may no longer need attention.

5 Leave prioritisation to organisational politics. In practice, powerful individuals in an organisation are likely to exert some influence even when one of the approaches above is applied. Indeed, any formal approach may itself be wielded as a weapon of power by those whom it favours.

Prioritising defects

An alternative way to prioritise SPC implementation is to focus on non-conformities instead of processes. This requires data to be collected on rejects and defects that have occurred over a period of time. The check sheet used to collect these data is regarded as a tool of SPC. It is nothing more than a table of counts of the types of non-conformity that occurred in successive time periods.

ACTIVITY 3.23

Consider for a moment whether counts of rejects and defects are available from the processes of your organisation. Do you think that they would be available in an organisation that has just begun to prioritise processes for SPC implementation? ●

It is often easier to interpret numbers from a check sheet if they are portrayed graphically, and one tool for this job is the Pareto chart, which you have already come across in Block 2 and earlier in this block. A Pareto chart representing defects in descending order of their frequency of occurrence will help to identify where remedial action is most needed.

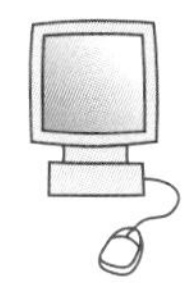

Now do Exercise 3.5 in the Computer Exercises Booklet.

However, a Pareto analysis does not include information about time. This omission can make an analysis very misleading, because the frequencies with which defects occur can change as time passes. Thus the older observations in a check sheet may include problems that were resolved long ago. To examine the relative importance of different defects over time, you can instead plot what is known as an area graph.

Now do Exercise 3.6 in the Computer Exercises Booklet.

ACTIVITY 3.24

Use of Pareto charts and area graphs prioritises the most frequent defects instead of the most critical processes as described earlier. Suggest two difficulties with prioritising frequent defects. ●

11.2 Measuring processes

A process will have many characteristics and outcomes, and deciding which should be quantified into data for SPC and how they should be measured is not easy. This section addresses these questions.

In Programme 1 the police measured their process in terms of the time to respond to urgent calls. This choice of metric seems straightforward because response time was a government-imposed target. You may have noted nonetheless that they analysed the response time into several components, only one of which was used to measure the process in question. They were also careful to define precisely what was urgent. The same sorts of issue must be addressed for any process:

- which process characteristics to monitor, given the particular reasons for using SPC
- which metrics to use to quantify the characteristics, given how the process works
- how to define the metrics so that measurements can be consistent and precise.

Process characteristics

It would usually be wasteful to apply statistical control to all except the few most important characteristics of a process. Which characteristics are important depends on the reasons for applying statistical control. The standard definition of SPC puts forward these general objectives:

> a) to increase knowledge about the process;
>
> b) to steer a process to behave in the desired way;
>
> c) to reduce variation of final-product parameters, or in other ways improve performance of a process.
>
> ... The common economic objective of statistical process control is to increase *good* process outputs produced for a given amount of resource inputs.
>
> (BS ISO 11462-1:2001, p. 1)

ACTIVITY 3.25 ....

Which of these general objectives were the main focus of the engineering firm, the police force, and the consultant neuro-physiologist as presented in Programme 1? ●

The neuro-physiologist was interested in several process characteristics, for example the amplitude, latency and velocity of nerve signals, as well as characteristics of the measuring equipment and the nerve being measured. It is reasonable to expect that the same would be true of anyone seeking to understand a process. Both the other cases focused on one process output, in one case for control and in the other case to illustrate process improvement. Clearly, the choice of what to measure is influenced by the measurer's goals as well as by the process.

Measures of output are certainly valuable when estimating process criticality in order to determine priorities. However, they may not be the best characteristics when it comes to controlling the process. They have two drawbacks. The first is that they can be monitored only after the end of the process: the ISO standard (BS ISO 11462-1:2001) calls them 'final-product parameters'. This is too late for a non-conformity to be prevented or corrected and possibly too late to prevent further non-conformities from the same cause. The second is that they represent the exterior of the process. This is of little help in diagnosing problems.

To overcome this problem, the ISO standard identifies two other types of characteristic: process parameters and in-process product parameters. A process parameter is a variable that describes or affects the input to a process, such as a machine operating temperature or a processing time. An in-process product parameter is an intermediate characteristic of the unfinished product, such as a size before machining or the number of errors fixed during an interim check.

Some care is needed in the choice of parameter. As the ISO standard explains, 'SPC operates most efficiently by controlling variation of a process parameter or an in-process product parameter that is correlated with a final-product parameter; and/or by increasing the process's robustness against this variation.' The important word here is 'correlated'. It means that, to be useful, metrics of input or process should have been shown to be good predictors of output.

It also helps to be aware of who chose a characteristic and why. This helps to identify limits on alternative metrics, and perhaps to assess whether better choices might be sought. Table 3.16 shows some of the more usual reasons and their origins.

Table 3.16 Reasons for selecting metrics

Customer's choice	Externally imposed	Organisation's choice	Measurer's choice	Other
Specified output	Mandatory target	Problem under investigation or suspicion	Already collected	
Inferred need	Standard metric of the organisation's sector	Traditional metric of the organisation	Easy to collect	
	Measurement system built in by equipment suppliers	Quality target Cost target		

In Programme 1 the engineering firm measured a specified output characteristic, whereas the police measured a mandatory target imposed by the Home Office. Neither had much option. In contrast the neuro-physiologist falls into the 'Other' category, as he chose a range of metrics that might provide information about the process he wanted to understand.

Metrics

Metrics are the counts and measurements that quantify process characteristics and hence provide the data values for statistical calculations. In SPC, counts are known as 'attributes data' and measurements as 'variables data'.

Attributes data record the presence or absence of a characteristic of each product or operation being examined. The simplest metrics are counts of this presence or absence, often counts of non-conforming items or 'defectives'. Proportions are an alternative to counts. A proportion is effectively a double count as it is a count of a characteristic divided by the total of all items examined. The police data in Programme 1 are proportions of conforming items, namely arrivals at incidents within the specified time.

Sometimes counts are made not of the presence or absence of a characteristic, but instead of the number of features that are present in one item. For many purposes this is a count of individual non-conformities or defects. Defects are not the same as defectives. A scratch and a hole are two defects, but when both are found on the same item there is only one defective. Moreover, depending on the item's specification, some defects (such as a scratch) may not render an item completely defective but merely reduce its quality.

Variables data consist of measurements, which are numerical evaluations of characteristics of each item being examined (for example diameter, breaking strength, delay) in terms of units on a continuous scale (for example length, force, time). Most of the data in Programme 1 consist of measurements and are therefore variable data.

There can be a wide choice of metrics even for one particular characteristic. True, some kinds of process will constrain what is possible, for example malfunctions in complex systems and surface blemishes are both much easier to count than to measure, but most will not. Generally speaking, counts are cheaper and easier to produce than measurements, because measurements require a more complex and precise collection system, and because one count may correspond to measurements of several variables. For similar reasons, counts of defectives may be preferable to counts of defects. However, measurements provide far more useful information for diagnosing problems. From a statistical perspective, many more counts than measurements are needed to get a reliable indication of trouble, partly because non-conforming items are counted only when a characteristic crosses a specification limit whereas the trend towards that limit may be obvious in measurements for some time in advance.

So there is a balance to be found between counts and measurements, and moreover the balance is dynamic. As time passes, organisations alter their views on which characteristics to examine and which are the best metrics of them. At an initial stage of SPC implementation, counts of final-product parameters may be preferable because they can be quickly implemented. The earlier parameters, and measurements, may have to wait until there is greater knowledge of what gives the best control and of how to collect data efficiently.

ACTIVITY 3.26 ....

If you find variation in the measurements of a process, to what might this variation be due? ●

Measurement variation obscures what is really happening in a process. At worst, it may render data too inaccurate to support SPC decisions. Therefore, before applying SPC to the process proper, it is necessary to ensure that the measurement of the process is itself capable, and thereafter measurement variation must be kept within acceptable limits.

Fortunately, there are statistical techniques for examining the performance of measurement systems, given appropriate data. The techniques are known as measurement systems analysis, and their use is sometimes called a gauge study, or a 'gage' study using the American spelling.

To give you an idea of the capabilities of a gauge study, I have prepared some data for you to analyse with the statistical software that accompanies the course. The data come from three operators taking two measurements each of 20 outputs of a process. Because each operator measured the same 20 outputs, any variation could be due only to the measurement system or to the conditions in which it worked, not to the process itself.

First you will look at a run chart, which displays measurements from the individual operators' measurements alongside the other, and thus enables them to be compared visually.

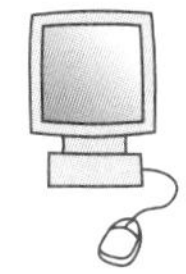

Now do the first part of Exercise 3.7 in the Computer Exercises Booklet.

ACTIVITY 3.27 ....

Where do you find greater variation in the results of the exercise – between different products or between different operators? What does this imply about the measurement system? ●

As is often the case, it is possible to confirm the impression you have from graphical results by calculation. In this instance, software can calculate the relative importance of the different components within the overall process variation, showing not only the influence of operators and products but also the interaction between them, for example particular operators having trouble with particular products.

Continue Exercise 3.7 in the Computer Exercises Booklet.

11.3 Bringing processes into control

Control has an exact meaning in SPC. A process is 'in control' when it exhibits only common-cause variation. Except for these random fluctuations, its behaviour is stable and predictable. Predictably, it will produce similar levels of defects today, tomorrow and thereafter. Even if the level of defects means that the process is incapable of meeting its specification limits, the process is nevertheless in control. The control chart that you generated from the parts data revealed just such a process.

Bringing a process into control, therefore, means identifying special causes of variation and preventing their recurrence or diminishing their effects, so that only common causes remain significant. The key tasks are to identify the point at which an out-of-control condition occurred, to diagnose what happened at that time, and to stop it happening again.

Detecting special causes

The primary tool for identifying out-of-control conditions is the control chart because it shows up such conditions as unusual patterns. The patterns are unusual because they are statistically very unlikely to happen by chance alone.

Return to the 'I chart of parts data' that you generated in Exercise 3.2 in the Computer Exercises Booklet. It contains the individual data points, connected to bring out their sequence, a line for the mean at 13, and two lines for the upper and lower control limits at 18.16 and 7.84. The control limits are calculated from the data, at three standard deviations (3 sigma, 3Σ or $3s$) from the mean; they are *not* decided by a person as specification limits are. Because of the way that the standard deviation is calculated as a measure of dispersion around the mean, only about 3 in 1000 points would fall more than 3Σ from the mean by chance, that is, due to random or common-cause variation. Therefore a point outside these lines on the control chart is likely to have a special cause.

A point outside the 3Σ lines on a control chart is in fact the most important of the unusual patterns that signal the potential presence of special-cause variation and therefore instability in the process. It is one of eight 'standard' patterns that have been defined by SPC practitioners over the years. SPC software detects these patterns automatically and highlights them when it draws a chart. So normally you do not need to hunt through control charts manually. However, such a hunt helps enormously when learning what the patterns are, so the next exercise asks you to do that.

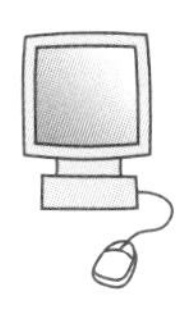

Now do Exercise 3.8 in the Computer Exercises Booklet.

Each distinct pattern is associated with different causes of variation. You will examine these later in this section, but first I want to look at the different types of control chart that are used to identify the patterns.

Selecting appropriate control charts

If you have explored the SPC software, you have probably noticed that it can produce several types of control chart. This section discusses how to choose between them.

The most fundamental choice is between attributes charts and variables charts. These types of chart relate to the types of process metric that you studied earlier – counts of attribute occurrences and measurements of variable quantities.

Attributes charts

SPC defines four types of attributes chart. Which one to use depends on what the count of attribute occurrences represents and on the sizes of the samples that were counted. Statistical calculations vary in each case. Table 3.17 lists the types of attributes chart.

Table 3.17 Types of attributes chart

Chart type	Use when
Defectives charts	
np-chart	Counts represent the *number* of defectives in each sample. The size of every sample must be the same.
p-chart	Counts represent the *proportion* of defectives in each sample. Sample sizes may vary.
Defects charts	Counts represent the number of defects in each sample.
c-chart	The size of every sample is the same.
u-chart	Sample sizes vary.

When defects occur rarely, their counts are too low for attributes charts to be practical. One solution that is often recommended is to measure the time between events instead. This is a good example of the creativity that can be applied in SPC. It also leads in to discussion of the other main type of control chart – the variables chart – which plots measurements instead of counts.

Variables charts

When the data represent measurements of a variable quantity, SPC plots two charts rather than one. It does this in order to examine separately two aspects of the measurements that can vary independently of one another: location and dispersion. The location is the central value. In the great majority of cases, the central value is deemed to be the average (usually the mean) of the measurements. In a few circumstances, a different expression of central tendency such as the median (middle) value or the mode (that is, the value that occurs most often) might be used, but you need not be concerned with them in this section. Dispersion is the spread of the measurements around the location, as represented for example by the width of the histogram that you found in the parts data when you did Exercise 3.3 in the Computer Exercises Booklet. Two expressions of dispersion are used in different circumstances, and you need to understand both. One is the range, which is simply the difference between the largest and smallest values. The other is the standard deviation. The standard deviation is less influenced than the range by extreme values.

The primary consideration in deciding which pair of charts to use is whether or not the data contain natural groups (sometimes called subgroups). There were no groups in Programme 1's example data of engineered parts or police responses. In contrast, the data in the measurement systems analysis example could be grouped by operator, product or both. It would be more usual, however, to use groups that are ordered in a time sequence. Groups in the data may represent a sequence inherent in the process, such as consecutive work shifts, or they may be created by sampling at time intervals, such as a group of five random observations taken each hour from the throughput of an office.

When there are groups in the data:

- The points on the chart of location are the means of each group in sequence. The chart is known as the $\overline{X}$ chart (say 'X bar'), from the statistical notation for the mean. A simpler but less used name is the average chart.
- The points on the chart of dispersion are either the ranges (R) or the standard deviations (s) of each group, in an identical sequence to the $\overline{X}$ chart. The chart is known as either the R chart or the s chart.
- Pairings of these charts occur so often that it is common to find them named as one chart such as $\overline{X}/s$ chart or X bar-R chart.

In the days before software, R charts were often preferred to s charts because they were easier to calculate. Now that both are easily produced, the s chart is a better choice in general because, as noted earlier, it is more robust to extreme values. It should certainly always be used when the sample size varies from group to group or exceeds 9. There is a caveat, which is that the R chart can conceivably identify an out-of-control point that both the $\overline{X}$ and s charts miss, when just one measurement in a group exposes a problem.

When there are no logical groups in the data:

- The natural sample size is 1. Consequently, there is no such thing as a sample mean, range or standard deviation. Instead, control charts use the mean and standard deviation of the whole set of measurements to determine where control limits are placed.

- The points on the chart of location represent the individual measurements, so it is known as an individuals chart or I chart (or in some quarters, confusingly, an *X* chart).
- The points on the chart of dispersion represent an artificially generated range, a 'moving range', calculated simply as the difference between the current measurement and the previous one. This is known as an MR chart. The pairing is known as an I-MR chart.

You have already produced an individuals chart in Exercise 3.2 in the Computer Exercises Booklet. Now it should be possible for you to make sense of the other types of variables control charts offered by the software. To do so, repeat that exercise and then explore the other menu options using the same data. Use the online Help to understand the options better.

Time-weighted charts

When a process has been in control, it might be useful to have an early warning if this changes, before it leads to non-conformities. The control charts that you have studied so far are not ideally suited to this. Process changes tend to be obscured by noise – the random movements of common-cause variation – and become obvious only after several observations. A better choice of chart for early warnings is a time-weighted control chart. In such charts noise is reduced, and also changes in the process are indicated much more clearly by *changes in the slope* of the line. This enables easier and earlier identification of the points at which changes occur.

Of the types of time-weighted chart provided by software, the two to focus on are the EWMA (exponentially weighted moving average) chart and the cusum (cumulative sum) chart. They are calculated from the same data as the charts you have already studied, but they analyse it in a different way. Do not confuse the EWMA with the ordinary moving average, a different type of chart that is seldom if ever used in SPC because it can be completely misleading.

EWMA charts

The line on an EWMA chart is a special kind of average that gives most weight to the latest observation and progressively less weight to all preceding ones. This has the effect of smoothing out the noise from random fluctuations, while emphasising the essential shape of the time series (which an ordinary moving average does not do).

Now do Exercise 3.9 in the Computer Exercises Booklet.

Cusum charts

'Cusum' is an abbreviation of 'cumulative sum', which refers to a procedure whereby a predetermined reference value, or target, is subtracted from each observation and the results are accumulated into a running total – the cumulative sum of the name. You will be familiar with this procedure if you have even a passing acquaintance with the game of golf, which is often scored by accumulating the number of strokes made above or below 'par', a target for each particular golf course, rather than by the total strokes made.

In the case of golf, the main function of par scoring is to enable comparison of players who have reached different points of the course, both against each other and against a historic standard of attainment. In process control, cusums can be used in the same way, for example to monitor attainment against target where output is gradual and variable, such as when testing a complex system. However, SPC in particular restricts itself to a usage seldom seen in golf, which is to reveal significant changes in attainment. This relies on a helpful effect that occurs when the reference value is held constant and the successive running totals are plotted on a cusum chart.

In a cusum chart, when an observation is above the reference value, the amount added is positive, so the cumulative sum increases and hence the line on the cusum chart slopes upwards. The reverse is true when a value is below the reference value. Thus any slope reveals whether a process is, on average, performing above or below its reference value, and any change in a slope reveals immediately when the level of performance changes. In consequence, a cusum chart can be even more sensitive to process changes than an EWMA chart.

You should be careful not to interpret the slope of a cusum chart as meaning the same as a slope in other charts. It does not indicate a changing value of the characteristic being plotted. Instead it indicates whether the average value is higher than, equal to or lower than the reference value (slope up, horizontal or down respectively).

Now do the first part of Exercise 3.10 in the Computer Exercises Booklet.

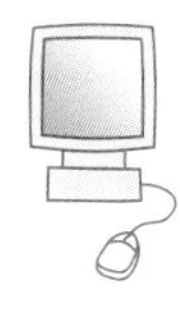

You probably noticed that two lines were plotted on the cusum chart, upper and lower, although the police data make the lower line almost flat. The upper line responds to processes above the reference value, while the lower line responds to processes below the reference value. This is known as a two-sided cusum chart. The sides are set apart because in some contexts (as in the police example) only one side is of concern, while in other contexts the sides can have distinct meanings. Historically, SPC also used a chart containing one combined line and V-shaped control limits, often called a 'V-mask' chart. More recent SPC texts (for example Montgomery, 2005, p. 405) strongly advise against the V-mask procedure because it is less informative and less sensitive to process changes. The police data support this view. Therefore you need not study V-mask charts for this course. However, you may come across them if you pursue SPC further and you can look in cusum options in the course statistical software for more information about V-masks.

The most important skill in cusum charting is choosing a suitable reference value, usually such that a flat line indicates an unproblematic process. This is because the easiest changes to detect are changes from the horizontal. When a line has an in-built slope, as happens when the process mean is consistently above or below the reference value, it becomes harder to distinguish significant changes.

ACTIVITY 3.28

What do you think might cause the process mean to be consistently different from the chosen reference value and lead to a persistent slope of the cusum? Suggest an instance where a slope might be useful. ●

For control purposes and for historical study, the best reference value is a suitable process mean. This raises a question of the period over which the mean should be calculated. For historical study, the entire period is a sensible choice. For control purposes, the mean should be in the vicinity of recent or expected (not desired) performance. This is why I used the mean of stage 3 in the software exercise above. An exact value, not rounded or approximated, is preferable, because it most emphasises changes from the horizontal.

Continue Exercise 3.10 in the Computer Exercises Booklet.

On the evidence here, cusum charts are preferable to EWMA charts. They can reduce background noise in the data and identify the precise point at which a process changes, without requiring a trade-off in effectiveness between these two abilities. They are thus suitable for monitoring and control, diagnosis and even short-term prediction. On the other hand, it is so easy for software to generate and fine-tune charts that cusum and EWMA charts are perhaps better regarded as companions rather than alternatives. If there is a discrepancy between them, this in itself is interesting information, and it would be useful to think carefully about what it might signify and what caused it.

Responding to out-of-control conditions

Once control charts can be produced regularly, three generic choices of action are possible:

1. Do nothing, because the process is in control and effort would be better spent on other processes.
2. Improve the process, because although it may be in control it is not as capable as it should be (see Section 11.4).
3. Respond to out-of-control conditions in the charts.

This section is about the third choice. I shall look at the other choices later.

Identifying what happened

Table 3.18 summarises the different signals that can be detected in control charts, alongside the standard tests for unusual patterns (numbered) that may reveal them. The tests are numbered after a convention used widely in SPC. The table also shows the most likely causes of these signals.

Table 3.18 Signals and their possible causes

Signal	Test	Possible causes
Single point outside the control limits followed by a return to normal	1	Measurement or recording error A temporary event such as a stand-in person A change in one of multiple sources of data, especially in a chart of dispersion
Step change in values	2, 5, 6	A sudden change in the process (e.g. damaged equipment; new staff, manager, procedure, supplier, target, bonus, etc.)
Gradual change in values	3, 6	Wear and tear on machinery or measuring equipment; relaxation of procedures
Alternating from high to low and back	4	The outcome of more than one process is being monitored, e.g. two shifts, machines, operators, suppliers Tampering when the process is in control, adjusting down after every high value and up after every low
Repeating pattern		Same as alternating, but with more than two components in the process Time of day, day of week, end of week/month/year effects. Always worth checking even if not obvious. [Repeating patterns are usually found only by personal inspection]
Many points near specification limits, but few over		Measurements being falsified to meet targets [This pattern is also found by inspection. Specification limits seldom coincide with calculated Σ lines. For assistance, you can add extra reference lines to the chart before or after it is generated; good SPC software provides both options]
Substantially more than 2/3 of the data within 1Σ of the mean	7	Measurements from different processes being added or averaged, thereby removing extreme values Data being manipulated to avoid recording extreme values
Substantially less than 2/3 of the data within 1Σ of the mean	8	Measurements from different groups (e.g. shifts, suppliers) with very different averages being mixed Tampering with the process, as in test 4

(Source: adapted from Stapenhurst, 2005, p. 261)

ACTIVITY 3.29

Several of the possible causes in Table 3.18 represent false positives in the data, such as a measurement error. Which causes do you think are likely to be genuine out-of-control conditions? ●

ACTIVITY 3.30

Figure 3.48 is an np-chart of the last two stages of the police data, with all tests switched on. ●

(a) What do the out-of-control points labelled '1' mean?

(b) What do you think they signify, taking account of the story in Programme 1?

(c) What do the out-of-control points labelled '4' mean?

(d) How would you explain them?

(e) How would you respond to the chart as it stands?

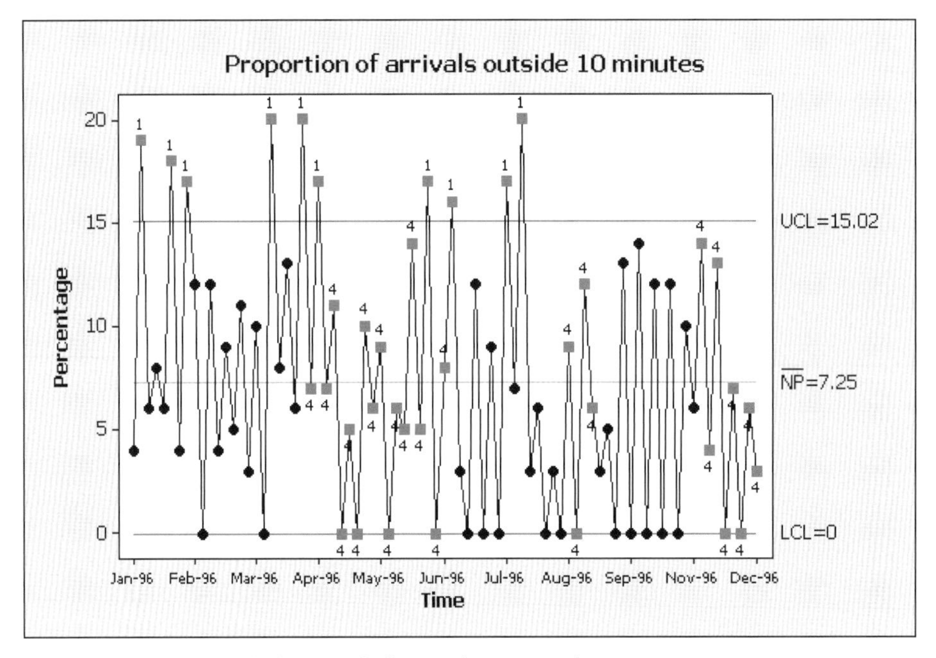

Figure 3.48 Control chart with flagged patterns for interpretation (Source: Minitab®)

Diagnosing causes

Process control charts are good for detecting problems but less good for diagnosing their causes. Unless you know which inputs and which conditions are significant for which behaviour or which outcomes, you have little more than guesswork with which to bring the process into control. Other tools are needed to acquire this knowledge.

Traditionally, SPC is credited with three tools for investigating causation:

- cause-and-effect diagram
- defect concentration diagram
- scatter diagram.

You have already met the first and third of these in Section 2. The defect concentration diagram presents a 'map' of a product or a process, annotated with the precise locations of defects, and sometimes quantities. As more data points are added to the map, patterns of locations are often revealed and give vital clues to the causes of problems. The diagram thus performs a similar role to a Pareto chart or a check sheet, but with additional information about spatial distribution.

The style of map varies according to the physical or logical architecture of the product or process. For example, a product might be mapped as a parts diagram, while a service might be mapped as a process flow chart. Montgomery (2005) uses refrigerator manufacture to provide the example shown in Figure 3.49. The finish was regularly damaged in the areas shown, first by a lifting belt round the middle and second by operators manoeuvring units on their front corners.

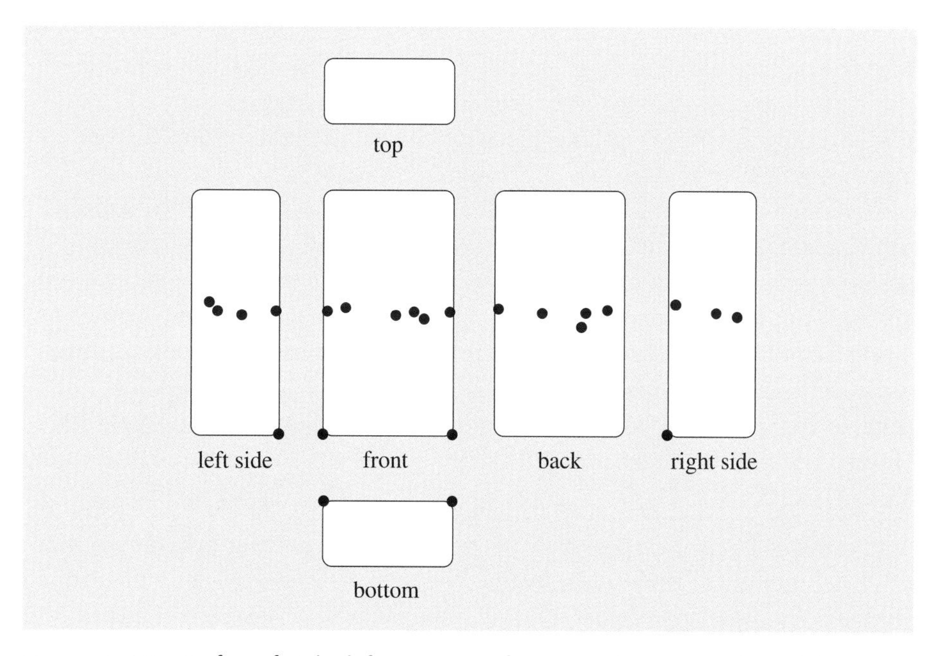

Figure 3.49 Surface finish defects on a refrigerator (Source: adapted from Montgomery, 2005, p. 174)

A defect concentration diagram is one of the most insightful SPC tools but also one of the least used. This is because it is not produced by software as readily as its companion tools; it must normally be maintained by hand on a preprinted map. Clearly the automated production of this type of diagram is beyond the scope of most SPC software. It might be worth adapting CAD (computer-aided design) or GIS (geographical information system) software to the purpose, if the need justifies it.

The scatter diagram, or scatter plot, which you studied in Section 2, examines a relationship between two variables. If the diagram describes an obvious pattern, there is a strong relationship between the variables. However, the scatter diagram and the linear correlation calculation often used with it are primitive tools suited only to the simplest situations. The statistical software on the course DVD provides a much wider range of graphs and analyses for examining relationships between variables. It is worth becoming familiar with these, even if you only browse the dialogues and related Help.

In a practical situation you are likely to be trying to understand the significance and interactions of several variables, not just two. Matrix plots display scatter diagrams of all the pairings within a set of several variables, enabling the user to estimate quickly which are the strongest relationships. You can add a regression (best-fit) line or a smoother (trend) line to emphasise the relationship visually. When a scatter diagram suggests a curved relationship, you should use a fitted-line plot instead of linear regression to examine it.

Scatter diagrams are limited in the number of variables they can examine at once; but several further types of diagram, including contour plots, 3D scatter diagrams and 3D surface plots, can reveal the relationship between three variables rather than two. Beyond three variables, statisticians have created other multivariate analyses, which are made easy to use by statistical software. One in particular, analysis of variance (ANOVA), is of interest because, among the variables whose relationships are being investigated, only one needs to be a numerical variable. The rest may be categorical variables, that is, identifiers of objects such as operators, machines, sub-processes and work shifts. ANOVA thus makes it possible to detect whether specific patterns in the data are associated with specific conditions, for example low defect rates with particular methods.

All statistical techniques for investigating relationships should be viewed with caution. Even if a relationship is rock solid, significant at less than the 1% level perhaps, all that is known is that an association exists in the data. Nothing is known about what kind of association it is, and in particular about whether it reflects a causal relationship between the phenomena that the data represent. Statistics are a good clue for the detective, but bad evidence in court, so to speak. For instance, a strong relationship between two phenomena may stem in fact from a third, unidentified cause. In some practical situations this may not matter: for example, if you are looking for a predictor variable, perhaps a process parameter or in-process product parameter such as a number of queries or a quantity of waste, which shows trouble brewing before process outcomes can be measured. In such cases, a clear trend on the scatter diagram of outcome against predictor is very welcome, whatever its cause.

Perhaps the best technique for investigating causality by observation is experimental design. It systematically adjusts the controllable input factors of a process, to identify which of them are important to outcomes. Even experimental design has its limits, however. Adjusting inputs to see what happens has proved to be inappropriate for some processes, such as the nuclear reactor at Chernobyl. Process variability is as likely to be due to changes in environmental factors such as ambient air temperature and humidity, or changes in process infrastructure such as wear and tear on machines, as it is to controllable input factors.

Ultimately, therefore, diagnosis of process problems relies a lot on craft knowledge, which can be acquired only by hands-on experience of the process. This is why the operators of a process are so often its best diagnosticians. Programme 1 showed operators taking measurements and diagnosing an out-of-control condition with their first-line supervisors, and this is the norm for SPC. To do otherwise can be counterproductive, as the example in Box 3.4 illustrates.

BOX 3.4 OWNERSHIP OF CONTROL CHARTS

Milton Technologies, manufacturers of advanced optical components for telecommunications systems, had successfully implemented statistical process control. For a number of years operators had been responsible for measuring 'critical to quality' characteristics of their output, updating control charts manually, and bringing any special-cause variation to the attention of their supervisor. The system worked well.

However, it was felt by plant management that improved control could be achieved by automating the collection and analysis of the control data. Capital was invested in more sophisticated measuring and data-logging equipment and software and in an SPC analysis and reporting package. The installation of the equipment and the implementation of the software went well and, apart from one or two minor teething problems, the new system was soon up and running.

Within six months the use of SPC in the plant had all but disappeared and process control had worsened. This was felt to be due to several causes:

- *The time taken to identify and correct special-cause variation.* Production supervisors and managers did not always have time to look at the output from the new system. They often had other, more pressing matters that had to be dealt with. As a result, corrective actions were taken slowly. In many instances this made it difficult to identify the root cause of problems which, in turn, meant that problems recurred.

- *Process management.* The older, paper-based system in effect made the operators the process managers. It gave them a way to keep the process under close scrutiny. Furthermore, involvement in the correction of special-cause variation would have taught them valuable craft knowledge of the process, adding to their competence.
- *Process ownership.* Close control would also have given operators a sense of owning the process, such that its objectives became their objectives.

ACTIVITY 3.31

I take three lessons from this story: 'if it ain't broke, don't fix it'; a simple SPC system can be a better solution; and devolved ownership of SPC can be a better solution. Assess whether these principles would pose a serious problem for organisations that you know, in terms of clashes with organisational culture. ●

Taking control action

Control charts do not in themselves control anything. SPC techniques become effective only when action is taken on the basis of the information that the charts and numbers provide. Who takes this action? As with diagnosis, SPC regards the process operator as its best controller. For this to be possible, operators need to be trained in techniques and given codified knowledge to supplement their craft knowledge.

With this objective in mind, SPC practitioners have settled on a mechanism called the out-of-control action plan, or OCAP. An OCAP documents process knowledge. Specifically, it includes diagnostic knowledge to help determine the causes of out-of-control conditions and the actions to resolve them. It may also include additional guidance, such as how to deal with any non-conforming products and other outcomes that have resulted from the out-of-control condition, and how to operate while the condition persists. It can be argued that no control chart should be put into formal use without an OCAP to guide its users.

Among the purposes of an OCAP are:

- to delegate control to operators, giving them as much responsibility as possible
- to document responsibilities
- to maximise the use of best practice across teams and individuals.

An OCAP is a living document. Each out-of-control condition, diagnosis and adjustment to the process potentially creates new knowledge that can update the OCAP. Updating can be automated to some extent if the OCAP is integrated with the defect-recording system from which charts are generated. The OCAP can then contain up-to-date results from many of the tools described earlier, such as an area graph of defects and defect concentration diagrams. It should also contain a history of the known causes of defects, with cause-and-effect analyses.

Maintenance of the OCAP is a primary responsibility of an SPC practitioner. It is essentially a responsibility for managing the organisation's knowledge about its processes, and for presenting this knowledge in a form that meets its purposes.

11.4 Making processes capable

Persistent remedial work on the causes of out-of-control conditions will eventually bring a process into control. In this state, process outcomes will be stable and predictable within the limits imposed by common-cause variation. However, as Programme 1 illustrated, these limits may still be too tight for the process. At this point, SPC turns its attention to process capability.

Making processes capable involves the control of common causes of variation. It is important to do this after eliminating special causes because special causes create instability in a process. Instability obscures common causes and, if you do not understand what common causes are present, you will not know if a process is capable of conforming to specification and therefore if the process must be redesigned to make it capable.

Assessing process capability accurately

ACTIVITY 3.32

Earlier in this section (Exercise 3.4 in the Computer Exercises Booklet) you learned how to generate statistics to estimate the capability of a process to stay within specification limits. Which of these statistics would tell you whether a process is capable when it is in control? Explain your choice, and indicate what values indicate sufficient capability for you. ●

Capability statistics indicate the location and spread of data, but accurate assessment of their implications also requires consideration of the data's frequency distribution. The distribution governs estimates of defect rates, that is, the proportion of outcomes outside specification limits. It is easiest to think of the distribution as the shape of the histogram of the data, or of the curve fitted to it as illustrated in Exercise 3.3 in the Computer Exercises Booklet. In that exercise, the data happened to fit a normal distribution quite well, and the estimated defect rates were based on that assumption. In practice, data are not always so obliging. Distributions can take many shapes, which means that estimates assuming a normal distribution can be much too high or low. In these cases an alternative distribution that fits the data better must be found.

If the data represent counts, the distribution is likely to be far from normal, but the choice of an alternative distribution is simple. When defectives are being counted (as plotted on p-charts or np-charts), the most obvious distribution to use is the binomial. When defects are being counted (as plotted on u-charts or c-charts), a distribution to consider is the Poisson. You will find both of these in the software's capability analysis tools. These analyses cannot provide capability ratios as such, although the process Z statistic for the binomial distribution is a substitute for Cp. Nevertheless both analyses generate estimates of defect rates that can be compared with other processes.

It would be unusual in practice to monitor process capability with counts. Measurements provide richer and more accurate estimates. However, when they stray from a normal distribution, the choice of an alternative is more complicated. Either you can choose whichever one of an assortment of non-normal distributions best fits the data, or you can transform the data to fit a normal distribution and analyse those figures instead. This depends on whether you need the extra information that analysis of normal data can provide, as discussed below.

The extra information from analysis of normal data really matters only when the data contain groups, which is often the case in practice, such as when samples are taken from successive work shifts. Then it can be useful to separate variation into components, namely the variation *within* groups and the variation *between* groups, as distinguished in Figure 3.50.

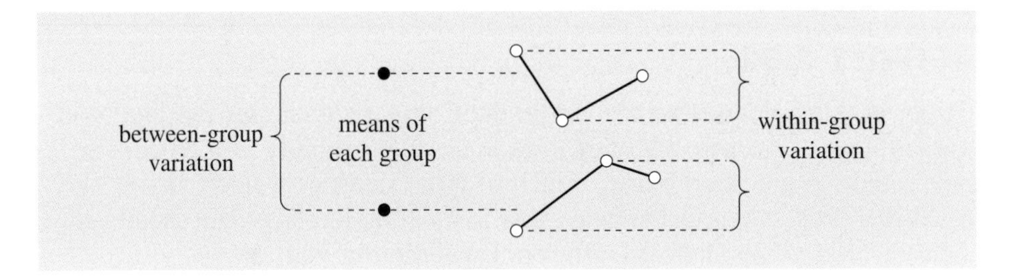

Figure 3.50 Between-group and within-group variation

The component variations may have different causes that can be addressed separately when improving the process. Analysis of normal data can generate statistics for both components, indicating where most of the variation occurs and hence where to look for the most significant causes. I suggest that you browse the examples in the software's Help system for both the between/within capability analysis and the normal analysis, so that you know which statistics they provide. You need both of them if you are to get the most information. The standard deviations in the between/within analysis tell you which component is relatively more important, while the expected performance figures in both analyses give an idea of the scale of any disparity.

If you need this extra information but are faced with a non-normal distribution, you must transform the data. Transformation calculates a new column of values for each data point, and then these new values are analysed in place of the old. For the analysis to be correct, any targets and specification limits that contribute to capability ratios must also be transformed in an identical manner. Transformation as presently implemented in the course software is complicated, and you do not need to study it.

Addressing low process capability

To recap, statistical analysis reveals different aspects of low process capability:

1. Variation may be too wide. This is shown when Cp (or Pp) is lower than the accepted norm, namely 1.33 in most cases or 2.00 in organisations pursuing Six Sigma goals, or when expected performance in terms of the number of non-conforming items per million is not acceptable.
2. The process may be performing off-centre between its specification limits. This is shown when Cpk is lower than Cp (or Ppk is lower than Pp), or when expected performance beyond one limit differs significantly from expected performance beyond the other limit.
3. Variation between groups such as teams or machine clusters may be relatively high. This is shown when the expected performance within groups differs significantly from expected performance overall.

To prioritise these aspects of low capability, it is probably easiest to use the various estimates of performance, at least until more subtle signals can be picked up with experience.

Assuming a stable process, wide variation has two main implications:

1. There is too much random, common-cause variation for the process to avoid frequent non-conformities outside its specification limits.
2. If common-cause variation cannot be reduced, the specification limits are unreasonable at present.

Both of these are predominantly strategic issues. To improve the process, some of the causes of variation must be eliminated or avoided and, as common causes are inherent in the way that the process currently works, this means that the process needs to be rethought. This may lead to redesigning flows, procedures, practices, equipment, skills and other resources, or the conditions within which the process operates.

In contrast, if the process is off-centre, perhaps the most straightforward cause is some maladjustment in the set-up. Another possibility is that observation took place when the process was in the act of drifting out of control, having breached a specification limit but not yet a control limit; if this is the case, it might be confirmed by a time-weighted control chart. In both these instances, the problem and its typical solutions may well be operational in nature, similar to those associated with process instability.

On the other hand, if the process suffers from built-in bias, this is a strategic issue that demands a rethink of the process.

At this point, SPC suffers from the same limitation that it does with special causes of variation: it is not very good at the precise diagnosis of causes and solutions. As Montgomery puts it:

> Essentially, SPC is a **passive** statistical method: We watch the process and wait for some information that will lead to a useful change. However, if the process is in control, passive observation may not produce much useful information.
>
> (Montgomery, 2005, p. 550)

Other tools are required. Montgomery goes on to recommend experimental design. This active method performs systematic tests on the process, changing inputs and observing the effects on outputs. It produces more information than do SPC's tools, but even so it is still limited because in this context it must work with the given process design. To rethink a process, an approach such as Six Sigma seems more appropriate than any problem-solving *technique*, because it works at a higher level. Judge this for yourself in the sections on methodologies in the next block.

Managing capability

Capability ratios may also be applied to ongoing process management. They can act as 'headline' figures, because they condense many data into one number, useful perhaps for reporting progress in a long-term improvement programme. They enable process comparisons to be made throughout an organisation, because they have essentially the same meaning whatever the process. They can also be routinely measured to monitor changes in common-cause variation. This last usage provides an opportunity to revise some parts of this section.

ACTIVITY 3.33 ....

(a) In what circumstances would monitoring capability ratios be a useful addition to control charting?

(b) Explain whether you would use a true capability ratio or a performance ratio for routine monitoring.

(c) Give an example of the type of change that monitoring Ppk would highlight.

(d) Give an example of the type of change that monitoring Pp would highlight.

(e) Which ratio (or ratios) would you monitor?

(f) What would you be looking for?

(g) What graphs or charts would you produce to monitor the ratios?

(h) How would you show progress towards a capability target? ●

11.5 SPC in context

According to Mason and Antony:

> The following are the typical benefits that can be gained from the application of SPC:
>
> - reduction in wasted efforts and costs;
> - process improvement, greater output;
> - better consistency of process output;
> - improved operator information: when to and when not to take action;
> - a predictable process can be achieved;
> - a common language on performance of process for different people across departments;
> - SPC charts help distinguish special from common causes of variation;
> - variation reduction;
> - reputation for high quality products/service and thereby reduce customer complaints;
> - healthy market share or improved efficiency/effectiveness;
> - reduced quality costs;
> - reduced need for checking/inspection/testing efforts;
> - more efficient management, and better understanding of process;
> - reduction in time spent fire-fighting quality problems.
>
> (Mason and Antony, 2000, p. 235)

When you examine this list critically, you will find that some items are merely features of SPC that precede benefits, some are benefits of SPC implementation itself, others are benefits that depend on implementing something in addition to SPC, and the remainder depend also on a desirable response from the outside world after implementation.

Naturally, SPC has weaknesses too. I do not mean weaknesses in its features, such as the lack of information for detailed diagnosis. You have already seen that these weaknesses can be countered with complementary tools like ANOVA, and they do not amount to much because, after all, SPC is popular and used successfully worldwide. Rather I mean weaknesses in the SPC philosophy or approach that might make it an unsuitable choice for some situations.

In some processes, wide variation is expected and welcome. These processes are commonly driven by the behaviour of customers. For example, salesmen often respond to customer demands by offering tailored products. Whether this is a good thing or not depends on the strategy of their business. If the strategy is to sell tailored products, in order to secure enough orders, then production processes must be able to cater for the resulting variation. This means that some process characteristics will not be useful for statistical control because they are expected to vary. However, the variety of process characteristics is usually so wide that substitutes may be found.

Some processes, on the other hand, should operate on target rather than within tolerance limits. Indeed, authoritative figures such as the Japanese engineer Genichi Taguchi say this of all processes, reasoning that the total cost to society increases in proportion to the square of the deviation from target. This calls into question the wisdom of using SPC to promote operation within tolerances rather than operation exactly on target. In other words it questions capability analysis. It does not, I feel, question the use of SPC to monitor variation and its causes. One way to resolve the issue might be to narrow the tolerances, so that they are tight around a target; this is in effect the approach of the Six Sigma movement. Taguchi favours a preventive approach based on prior experimental design of robust processes. Numerous statisticians have heavily criticised his particular choice of design technique as unnecessarily large and unreliable, but with better alternative techniques his philosophy is widely supported. Nevertheless, prior design of robust processes is perhaps too idealistic an approach for organisations that have to move forward incrementally from current processes.

SPC is a tool that enables practitioners to systematically understand, improve and control processes of all kinds. It can address any problem that is influenced by variation in process behaviour, such as low product quality or service quality, low productivity, excess wastage and excess emissions.

Problem solving with SPC is founded on the quantification of process characteristics, statistical analysis of variation in the resulting data, and inference of solutions from the analysis. This section has covered all three topics. It has also introduced you to the software for the statistical analysis that is provided on the course DVD. Such software brings SPC within the reach of even the smallest organisation.

Once a problem has been solved with SPC, the same tools can also ensure that it remains solved, by applying operational control to an ongoing flow of data from the process. Such control is essential in competitive environments where quality requirements become ever more exacting, and is an important aid to effectiveness and efficiency in general.

12 IMPLEMENTATION

The techniques that will be covered in this section are:

- weighted score method
- decision trees
- solution selection matrix
- risk assessment
- a simple step procedure
- Gantt charts.

12.1 Weighted score method

The weighted score method is a way of deciding which of a series of different options should be implemented. The first step is to draw up an extensive list of factors or quality dimensions against which each option can be judged. Weights are then assigned to each factor to reflect its relative importance and the weights are scaled so that the sum of the weights equals 1. Each option is then scored against each factor and, usually using a spreadsheet, a weighted score is obtained by multiplying the score by the weighting. The total weighted score for each option is obtained by adding the weighted scores across all factors. An example where the technique was used to select a piece of equipment is shown in Table 3.19.

Table 3.19 Example of weighted score method

Factor	Importance	Option A		Option B		Option C	
		Score	Weighted score	Score	Weighted score	Score	Weighted score
Technical service	0.3	80	24.0	65	19.5	60	18.0
Purchase price	0.2	20	4.0	50	10.0	80	16.0
Reliability	0.2	80	16.0	60	12.0	40	8.0
Annual energy costs	0.1	50	5.0	60	6.0	40	4.0
Availability of spares	0.1	20	2.0	60	6.0	70	7.0
Projected life	0.1	75	7.5	40	4.0	65	6.5
Total weighted score			58.5		57.5		59.5

12.2 Decision trees

A decision tree provides a structure within which options and the possible outcomes of choosing those options can be set out very clearly. As you can see from Figure 3.51, a decision tree is based on a tree diagram with two types of node: decision nodes (shown as a square), from which lines representing each option's action branch out; and chance nodes (shown as circles), which branch out into lines that lead to each possible outcome of that action. The probability of each outcome occurring can be included, as in Figure 3.52, and in this case the sum of the probabilities on the branches from one chance node must equal 1.

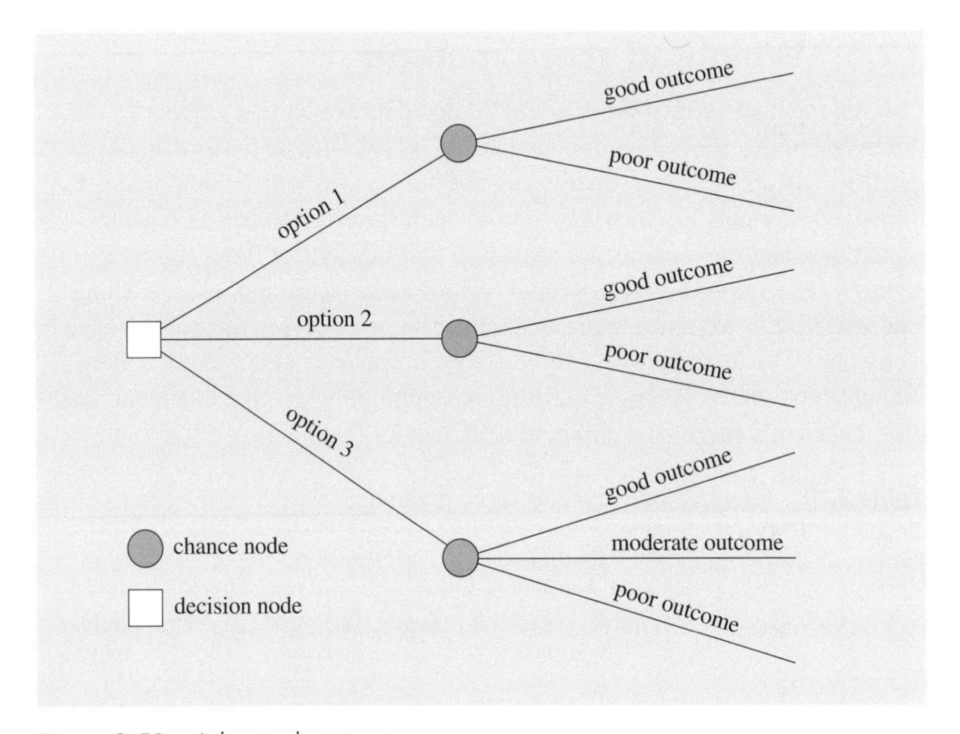

Figure 3.51 A basic decision tree

The first node of the tree must be a decision node, and the options form the first set of branches. The type of node that is at the end of each branch depends on whether the outcome from that option is certain or is determined by the outcome of chance events. This branching continues until the final outcome of taking each of these routes is reached.

If the purpose of the decision tree is just to show the options available then a basic tree showing the possible decisions after each node will usually contain sufficient information, but if the decision tree is going to be used to aid decision making it must be possible to estimate the values (V) of the outcomes. These values are usually expressed as cash values but they could be scores based on an alternative measure or set of measures. They are shown on the right-hand side of Figure 3.52.

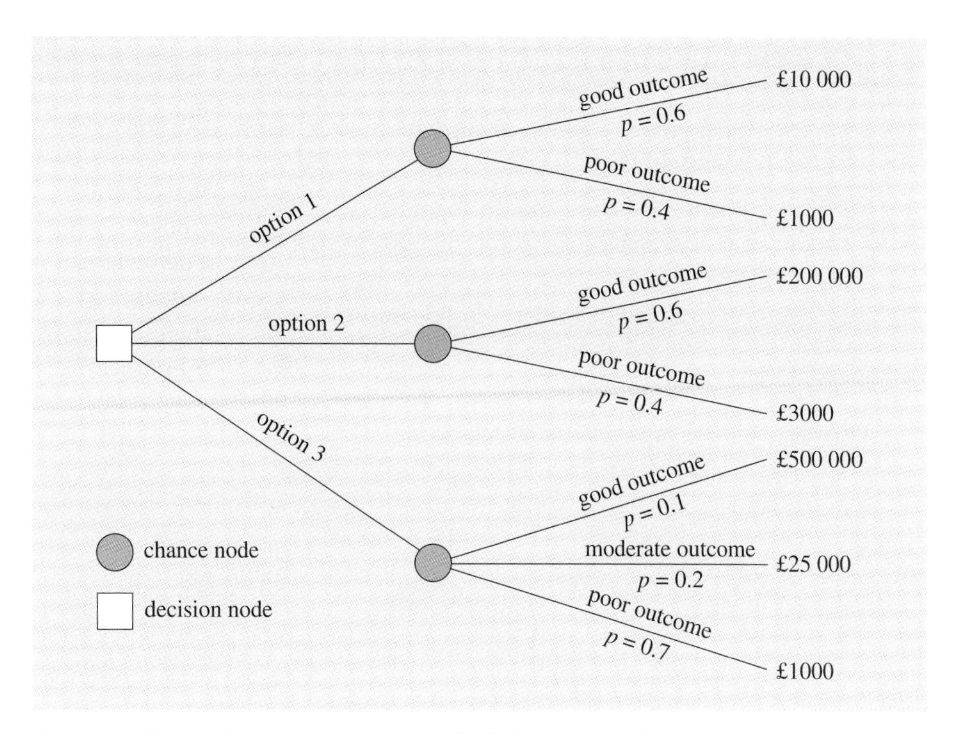

Figure 3.52 A decision tree with probabilities and values

The expected monetary value (EMV) for each option is given by:

$$EMV = \Sigma \text{ (probability} \times \text{value)}$$

So for Figure 3.52:

Option 1	EMV	$= 0.6 \times 10\,000 + 0.4 \times 1000$	= £ 6 400
Option 2	EMV	$= 0.6 \times 200\,000 + 0.4 \times 3000$	= £121 200
Option 3	EMV	$= 0.1 \times 500\,000 + 0.2 \times 25\,000 + 0.7 \times 1000$	= £ 55 700

The EMV for each option can then be compared with its cost. If we assume the costs of implementing options 1, 2 and 3 are £4000, £42 000 and £40 000 respectively, then the net benefits of the three options are as follows:

Option 1	6400 – 4000	= £ 2 400
Option 2	121 200 – 42 000	= £79 200
Option 3	55 700 – 40 000	= £15 700

Option 2 is therefore the most valuable even though it is the most expensive to implement.

Box 3.5 explains how a complex decision tree can be pruned back to a single decision node.

BOX 3.5 PRUNING A COMPLEX DECISION TREE

A complex decision tree can be pruned back to a single decision node by making two assumptions:

1. for every chance situation the decision maker is indifferent to whether it is represented by a certain sum of money or by the chance event;
2. the decision maker will prefer a large sum of money to a small one.

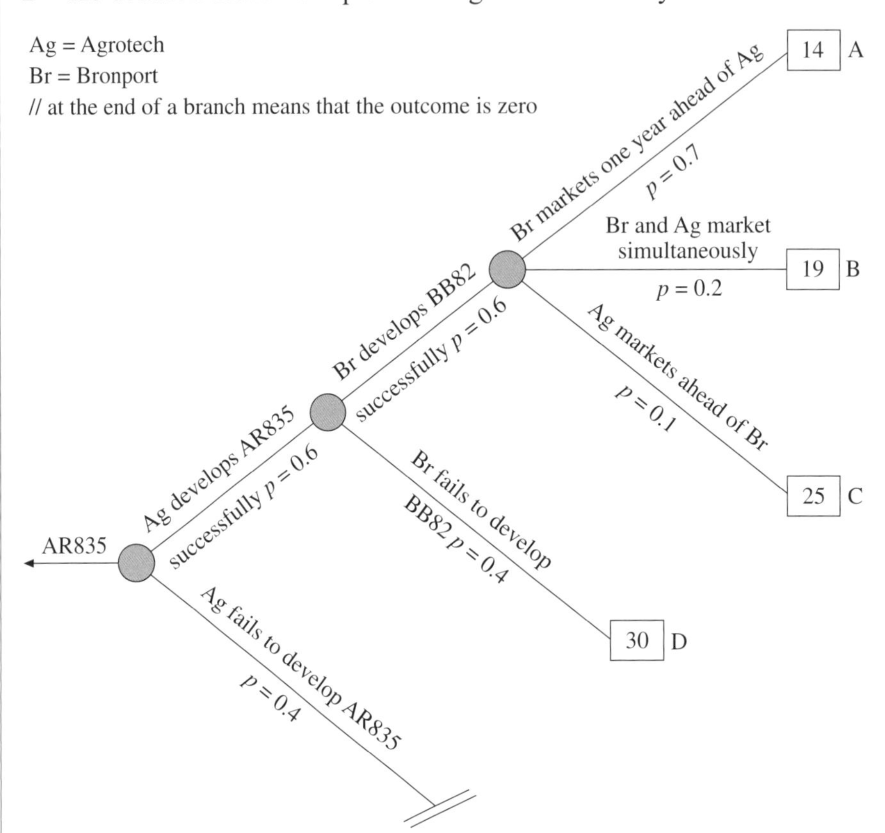

Figure 3.53 Section of a complex decision tree

The section of a complex decision tree shown in Figure 3.53 can be pruned back until you reach a simple tree comprising one set of branches. To roll back the branch of the tree shown in Figure 3.53, go to the furthest node and calculate the EMV:

$$\text{EMV} = 0.7 \times 14 + 0.2 \times 19 + 0.1 \times 25 = 9.8 + 3.8 + 2.5 = 16.1$$

This value of EMV replaces this furthest node.

Going back to the next remaining node:

$$EMV = 0.6 \times 16.1 + 0.4 \times 30 = 9.66 + 12 = 21.66$$

This value of EMV replaces this node.

Going back to the first node:

$$EMV = 0.6 \times 21.66 + 0.4 \times 0 = 12.996$$

ACTIVITY 3.34 ....

Prune back the section of a decision tree shown in Figure 3.54. ●

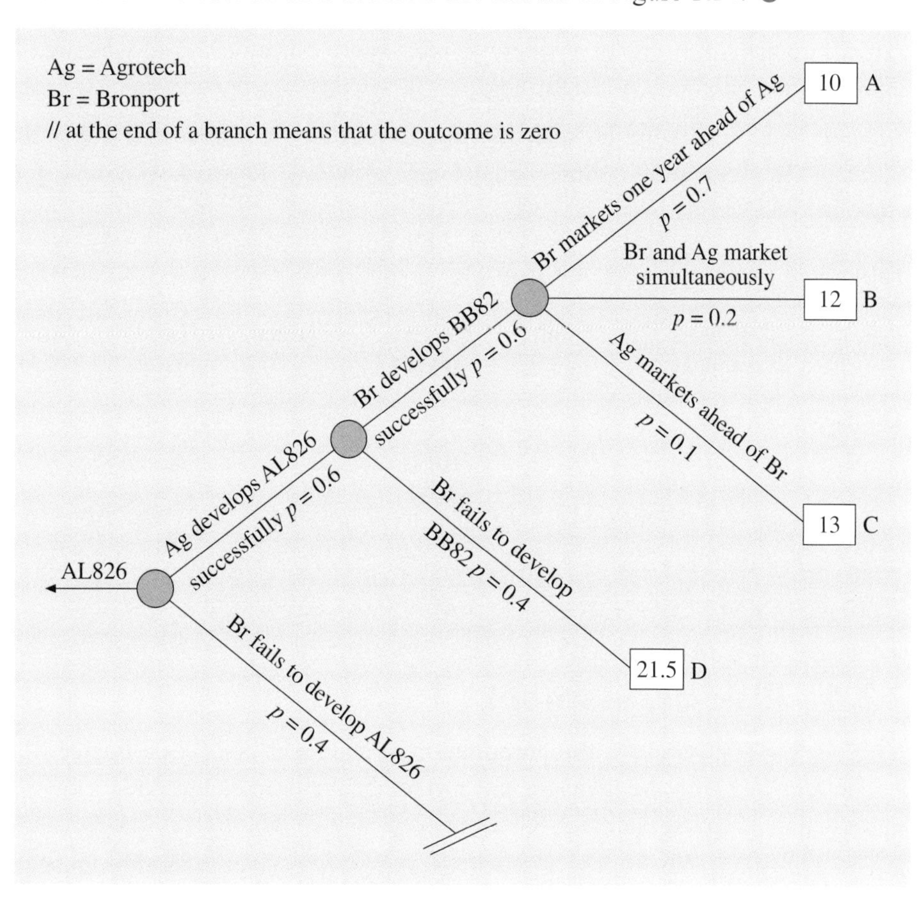

Figure 3.54 Decision tree to be pruned

12.3 Solution selection matrix

A solution selection matrix is another way of selecting which option to implement. The technique comprises a series of stages.

Stage 1. Remove 'showstoppers' from the list of options. 'Showstoppers' contain components that could not be implemented. For example, they may cause an option to cost too much or take too long or breach health and safety legislation.

Stage 2. Examine the organisational fit of each of the remaining options. This needs to take account of people factors, strategic factors and the operating and management systems that are already in place. For example, an improved method of working that requires people to work in multi-skilled teams may not fit well with a situation that has always rewarded individual effort. Although a good fit is desirable, lack of fit does not necessarily rule an option out. It does, however, signal the need for special measures to be taken in order to bring about a successful implementation.

Stage 3. Use the objectives that were identified at the start of the problem-solving or improvement exercise (see Section 3.2 of Block 1) to evaluate the extent to which each option is likely to meet those objectives. The weighted score method can be very useful here.

Stage 4. Draw up a solution selection matrix of the type shown in Table 3.20 and score the remaining options against each criterion. Again, a weighted score method can be used.

Table 3.20 Solution selection matrix

	Process impact	Time	Cost vs. benefit	Other	Total score	Rank
Weight	2	2	3	1		
Option 1	8	8	10.5	4	26.5	2
Option 2	14	18	22.5	7	32.5	1
Option 3	2	4	21.0	1	28	3

(Source George et al., 2005, p. 260)

12.4 Risk assessment

Because implementation is the introduction of change it involves a degree of risk in terms of both the management of the change process and the introduction of unfamiliar, and often untested, new structures, processes, procedures and products. Risk assessment is a way of identifying and quantifying those risks so that they can, where necessary, be avoided or be managed so as to minimise potential damage and disruption.

The process of risk assessment is shown in Figure 3.55. BS IEC 62198:2001 suggests the following methods for identifying hazards:

- brainstorming;
- expert opinion;
- structured interviews;
- questionnaires;
- checklists;
- historical data;

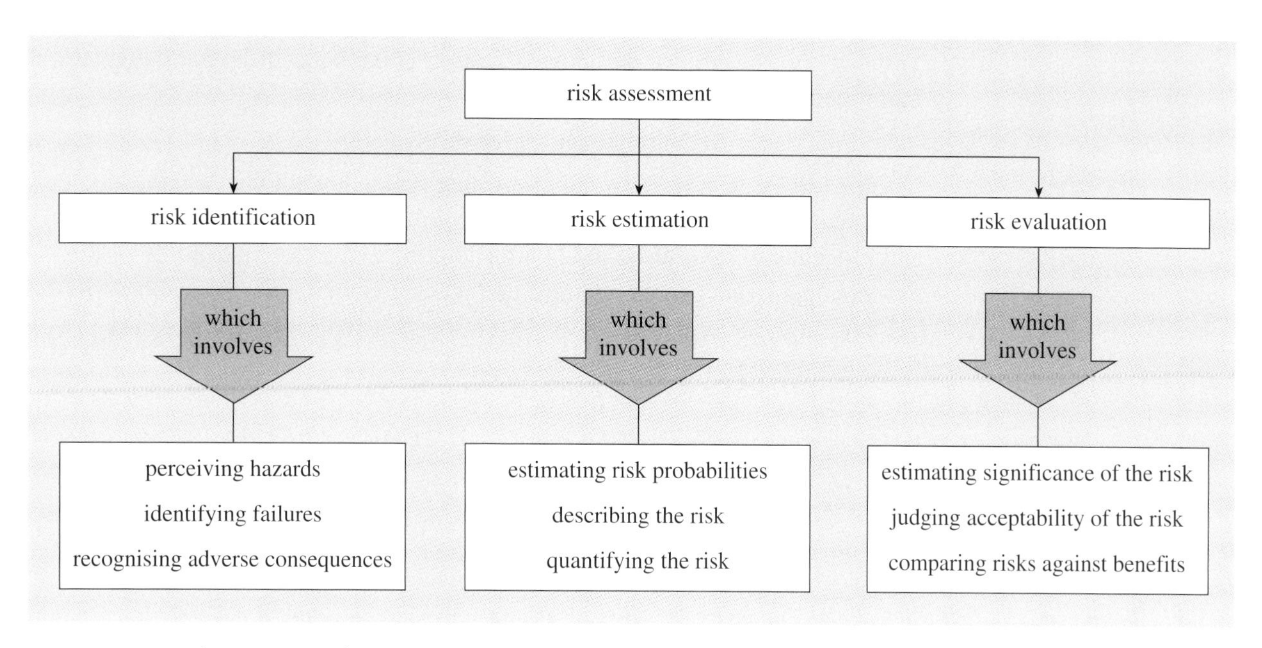

Figure 3.55 The process of risk assessment (Source: White, 1995, p. 36)

- previous experience;
- testing and modelling;
- evaluation of other projects.

(BS IEC 62198:2001, p. 12)

ACTIVITY 3.35

Of the techniques you have studied in this block, which three, other than the alternatives to brainstorming in Section 8, would you expect to be most useful in identifying potential risks? ●

In estimating risk probabilities it is usually sufficient to classify the probability as high, medium or low, although there may be occasions where it is possible to be more precise. For example, historical data may be available to show the likelihood of a delivery arriving late.

The concept of vulnerability is very important in relation to the estimation of the significance of risk. Vulnerability is the probability of being damaged by a hazardous event. For example, your vulnerability to a disruption in the supply chain will become much greater if you reduce the amount of stock that is held. Other factors that may need to be taken into account include the likely frequency of hazardous events, the extent to which they are predictable and the amount of advance notice of their onset and the duration of each occurrence.

The total risk faced in a particular situation is the sum of the consequences multiplied by their probabilities:

$$\text{total risk} = \Sigma(\text{consequence} \times \text{probability})$$

but this total sum is seldom the overriding factor in deciding whether to undertake a particular initiative. Risk has to be set against potential benefits, and judgements have to be made after looking across a spectrum of risks. For example, what account should be taken of events with low probability but very high consequence? Table 3.21 shows a typical example of the follow-on from a risk assessment.

Table 3.21 Sample summary table for safeguard implementation

(1) Risk (vulnerability/threat pair)	(2) Risk level	(3) Recommended controls	(4) Action priority	(5) Selected planned controls	(6) Required resources	(7) Responsible team/persons	(8) Startdate/end date	(9) Maintenance requirement/ comments
Unauthorized users can telnet to XYZ server and browse sensitive company files with the *guest* ID.	High	Disallow inbound telnet Disallow 'world' access to sensitive company files Disable the guest ID or assign difficult-to-guess password to the *guest* ID	High	Disallow inbound telnet Disallow 'world' access to sensitive company files Disabled the *guest* ID	10 hours to reconfigure and test the system	John Doe, XYZ server system administrator; Jim Smith, company firewall administrator	9-1-2001 to 9-2-2001	Perform periodic system security review and testing to ensure adequate security is provided for the XYZ server

(Source: Stoneburner et al., 2002, p. C-1)

12.5 A simple step procedure for implementation

Unfortunately, many problem-solving and improvement projects run out of steam as they approach the implementation phase. One very simple, but effective, procedure for implementation is based on the answers to the following questions:

1. What needs to be done?
2. In what order?
3. When?
4. By whom?
5. What resources are needed and how can they be secured?

12.6 Gantt charts

A very common and effective way of displaying a schedule of implementation activities is to use a bar chart such as that shown in Figure 3.56. Bar charts of this kind are usually called Gantt charts after Henry Gantt, who is credited with inventing them during the First World War. The activities are listed vertically and horizontal bars represent the period over which each activity is to be performed.

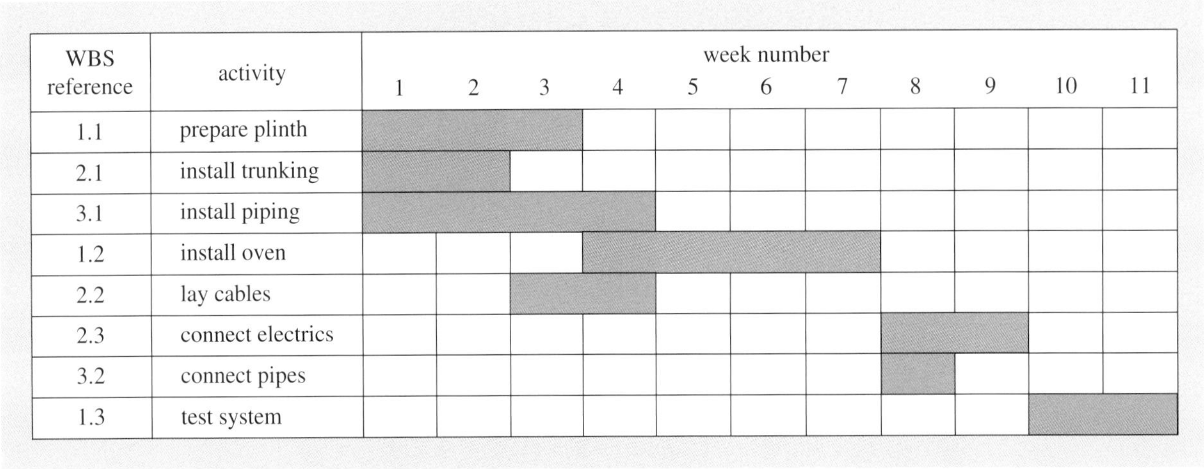

WBS reference	activity	week number										
		1	2	3	4	5	6	7	8	9	10	11
1.1	prepare plinth											
2.1	install trunking											
3.1	install piping											
1.2	install oven											
2.2	lay cables											
2.3	connect electrics											
3.2	connect pipes											
1.3	test system											

Figure 3.56 Gantt chart for installing a new industrial oven

The Gantt chart shown in Figure 3.56 is a very simple form of the convention. Its simplicity makes it particularly attractive as a way of conveying formation about the schedule to the people involved in the implementation. It is very easy to understand but the amount of content is fairly limited. There are many variations on the simple Gantt chart, which convey more information. For example, colour or shading may be used to show which activities are the responsibility of a particular person or department, or to indicate the current state of progress on each activity. Extra symbols may be inserted to highlight key events, such as the completion of the preparations to install the oven at the end of week 7 and the readiness for testing at the end of week 9. An example of a much more sophisticated Gantt chart is shown in Figure 3.57.

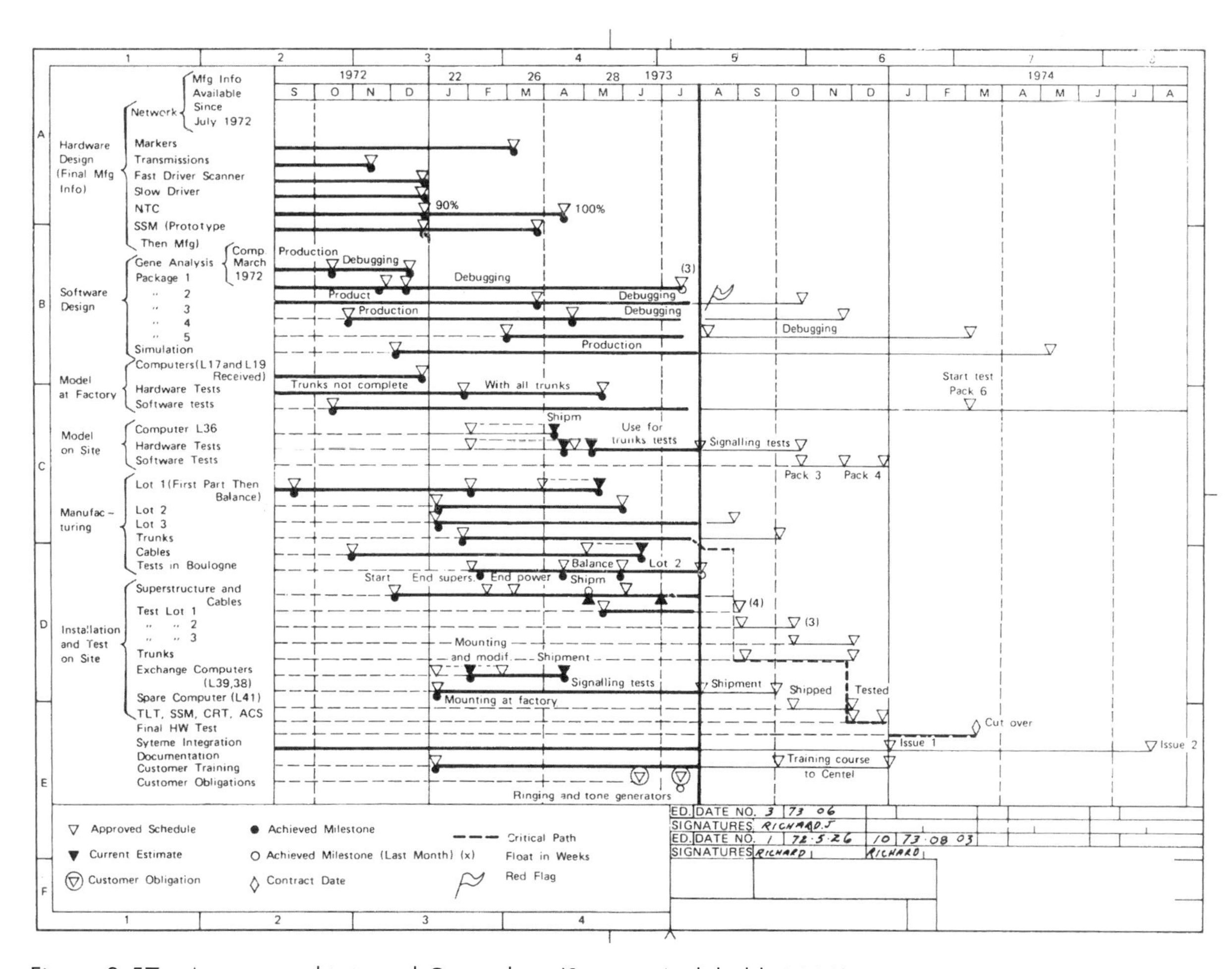

Figure 3.57 A more sophisticated Gantt chart (Source: Archibald, 1992)

One of the things a Gantt chart can do is highlight the chain of sequential activities that determines the minimum time required for the project. These activities are said to be on the critical path and delay in any one of them would lengthen the project. Critical path and the associated concept of float are very important. Some activities could increase in duration or be started late, yet the project could still be finished by the target date. Such activities are said to exhibit float. Float is the excess time available for an activity in addition to its estimated duration. The maximum time allowable for any activity may be calculated as the difference between the latest time by which the activity must finish and the earliest time at which it could start. Hence the total float for an activity is given by subtracting the estimated duration from this allowable time, giving the following formula:

total float = latest time of finish event
– earliest time of start event – duration

It is essential to realise that float is a resource that is *shared* between activities. If any activity uses any of the total float then the float of the other activities on the path is correspondingly reduced, so other activities have less room for slippage. The concept of float leads to the standard definition of a critical path: a critical path is a path with least float.

ACTIVITY 3.36

Will a delay in an activity on the critical path cause the implementation to take longer than intended and thus miss its required completion date? ●

The main weakness of the Gantt chart is that it is not easy to show the dependence of one activity on another, except by inference. For example, you can see in Figure 3.56 that the trunking is installed by the end of week 2 and that the activity of laying cables is scheduled to start at the beginning of week 3. This is unlikely to be a coincidence, so it can be inferred that one activity has been scheduled to follow the other. In a small project this may be obvious but in a large project there could have been many activities ending at week 2 and the inference that it is only the trunking that matters might be wrong. One way of overcoming this is to create lines linking activities that depend on one another. A drawback is that this tends to produce a rather cluttered chart.

13 FEEDBACK

The techniques that will be covered in this section are:

- key performance indicators
- radar diagrams
- balanced scorecard.

13.1 Key performance indicators

Key performance indicators are defined in Box 3.6.

BOX 3.6 KEY PERFORMANCE INDICATORS

A key performance indicator (KPI) helps a business **define** and **measure** progress toward its goals. KPIs are **quantifiable** measurements of the improvement in performing an activity that is critical to the success of a business.

KPIs should complement a business' overall targets and relate to its core activities. As a result, they will differ depending on the business. In a telesales business, for example, answering customer calls before they ring off will be a key business activity. The percentage of calls answered within one minute may be one of its KPIs.

There must be a way to define and measure KPIs if they are to be useful. For example, a business wanting to increase sales needs to consider whether to measure this by units sold or by value of sales, and whether to deduct returned goods from sales in the month of the sales or the month of the return. You also need to set a **target**, such as increasing sales by five per cent per year.

Financial KPIs focusing on sales, costs or working capital are popular as they enable businesses to monitor and control the profitability and cashflow of the business. For example, a KPI for monthly sales enquiries will warn you about peaks and troughs that will affect cashflow.

KPIs can also be used as a **performance management** and **improvement** tool by focusing your employees on achieving the business' goals. KPI monitoring enables management to spot and correct weaknesses in the business, for example in terms of cashflow.

(Source: Business Link, 2007)

Use of key performance indicators has very many supporters, especially in the public sector. For example, Table 3.22 shows the environmental key performance indicators recommended by Defra for a typical company. However, the approach is not without its critics. The biggest criticism picks up Deming's 11th point that you saw in Block 1:

> Point 11. Eliminate work standards that prescribe numerical quotas for the workforce and numerical goals for people in management. Substitute aids and helpful leadership; use statistical methods for continual improvement of quality and productivity.

Callaghan provides an example that concerns public transport in the UK. In setting 'quality of service targets' the Department of Transport introduced 'percentage of schedule operated' as one of its main measures of performance. However, as Callaghan pointed out:

> Managers ... have the ability to manipulate the service to increase their percentage of schedule, at the expense of making the service delivered to customers worse. For example, if the service were disrupted, and they were losing miles, managers on lines with more than one branch could route trains down the longer branch (which would increase mileage) and provide no service down the shorter. A true case of 'What gets measured gets done'.
>
> (Callaghan, 1992, p. 175)

Joiner and Scholtes cite the following entertaining example of the 'what gets measured gets done' phenomenon, although I am afraid I suspect it might be apocryphal:

> It is interesting to note that Management by Control is widely used in [the] Soviet Union. Typical is this story: Several years ago there was a surplus of large nails and a shortage of small ones. Why? Managers were held accountable for the tons of nails produced. Later the control was changed to the number of nails produced. This led to a shortage of large nails, since the smaller nails gave higher counts.
>
> (Joiner and Scholtes, 1985, p. 4)

Table 3.22 Environmental key performance indicators – Financial Year 2005

Direct impacts (operational)								
Greenhouse gases	**Definition**	**Data source and calculation methods**	**Quantity**					
			Absolute tonnes CO_2		**Normalised tonnes CO_2 per £M turnover**			
			2004	**2005**	**2004**	**Target**	**2005**	**Target**
Gas	Emissions from utility boilers	Yearly consumption in kWh collected from fuel bills, converted according to Defra Guidelines						
Vehicle fuel	Petrol and diesel used by staff and van hire fleet	Expense claims and MOT recorded mileage, converted according to Defra Guidelines						
Waste	**Definition**	**Data source and calculation methods**	**Quantity**					
			Absolute tonnes		**Normalised tonnes waste per £M turnover**			
			2004	**2005**	**2004**	**Target**	**2005**	**Target**
Landfill	General office waste, which includes a mixture of paper, card, wood, plastics and metals	Volume of waste generated per annum, calculated by recording the number of bins and skips removed, converted to tonnes according to Defra Guidelines						
Recycled	General office waste recycled, primarily cardboard	Volume of waste recycled per annum, calculated by recording the number of bins and skips removed for recycling, converted to tonnes according to Defra Guidelines						

Indirect impacts (supply chain)

Greenhouse gases	**Definition**	**Data source and calculation methods**	**Quantity**					
			Absolute tonnes CO_2		**Normalised tonnes CO_2 per £M turnover**			
			2004	**2005**	**2004**	**Target**	**2005**	**Target**
Energy use	Directly purchased electricity, which generates Greenhouse Gases, including CO2 emissions	Yearly consumption of directly purchased electricity in kWh, converted according to Defra Guidelines						
Water	**Definition**	**Data source and calculation methods**	**Quantity**					
			Absolute cubic metres		**Normalised cubic metres water per £M turnover**			
			2004	**2005**	**2004**	**Target**	**2005**	**Target**
Supplied water	Consumption of piped water. No water directly abstracted by the Group	Yearly consumption of purchased water						

(Source: Defra, 2006, p. 27)

13.2 Radar diagrams

Radar diagrams, also known as polar diagrams, are very useful for displaying performance indicator information. They allow a visual comparison between two or more situations, organisations, products, or whatever, across several quantitative or qualitative measures. Starting from a central point, each aspect is represented by a line radiating out from that point. Points are plotted on each radius for each comparator, and the points in each set are joined to generate shapes that can be compared visually.

Figure 3.58 is a simple diagram where the performance of one police force is compared with average performance across all police forces. Figure 3.59 is a more complicated diagram where five different staple crops have been rated according to eight criteria (1 represents the bottom rating and 10 represents the top).

13.3 Balanced scorecard

The balanced scorecard approach is described in an offprint.

Now read Offprint 6.

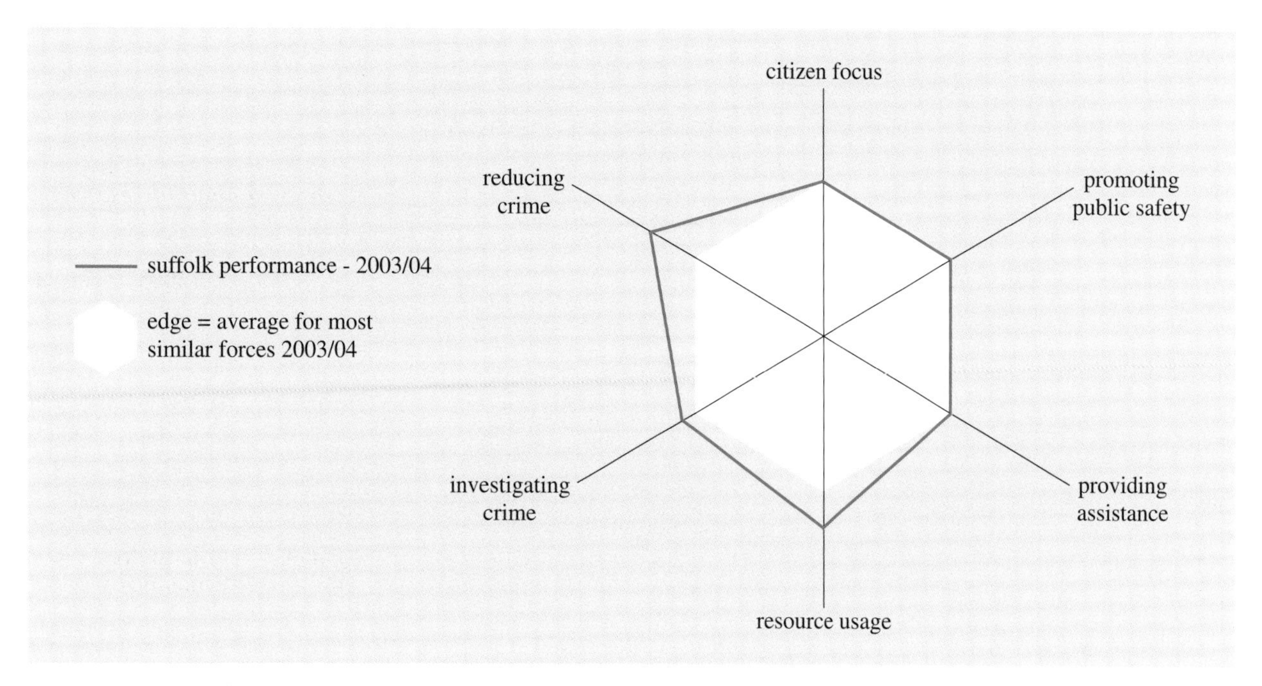

Figure 3.58 Radar diagram showing the performance of Suffolk Police in 2003/2004 (Source: redrawn from Police Standards Unit, 2004, p. 87)

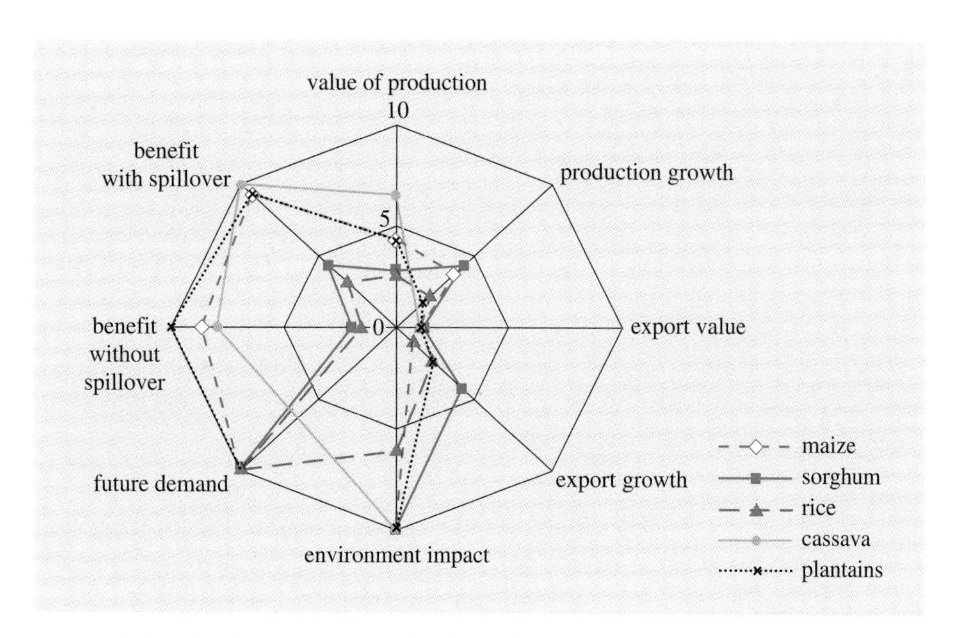

Figure 3.59 Radar diagram for food staples (Source: You et al., 2004)

14 CONCLUSION

This block has contained a vast array of techniques. Even used individually, each of them can contribute to problem solving and improvement by building awareness, generating understanding, and helping to develop and implement solutions or improvements and to monitor their effects. Used in combination in an organised way, they are even more powerful. In the next block I shall look at problem-solving methods and approaches that provide that organisation.

Before you move on to Block 4, however, I should like you to cast your mind back to the start of this block where I asked you to complete a table to show which parts of the DRIVE model you felt you could be satisfied by each of the techniques in this block. My (non-definitive) version follows.

Table 3.23 Techniques for defining, reviewing, investigating, verifying and executing

Technique	Define	Review	Investigate	Verify	Execute
Pareto analysis	X	X			
Cause-and-effect diagrams	X				
Stratification		X		X	
Tally cards		X	X	X	
Histograms	X	X	X		
Scatter diagrams		X		X	
Shewhart control charts		X	X	X	
Affinity diagrams	X				
Relational diagrams	X				
Tree diagrams	X	X			
Matrix diagrams	X				
Program decision process charts	X	X			
Arrow diagrams	X				
Matrix data analysis		X	X		
Is/is not analysis	X			X	
Five whys	X	X	X		
Five Ws and H	X	X	X		
More specific sets of questions	X	X	X		
Input-output diagrams	X				
Systems maps	X				
Influence diagrams	X				
Rich pictures	X				
Activity sequence diagrams	X				
Flow-block diagrams	X	X	X		
Flow-process diagrams	X	X	X		
Spaghetti charts	X		X		
SIPOC charts	X				
Multiple-cause diagrams	X	X			
Force field analysis	X				
Cognitive mapping	X				

Technique	Define	Review	Investigate	Verify	Execute
Failure modes and effects analysis			X		
Fault tree analysis			X		
Brainstorming	X	X	X		
Brainwriting	X	X	X		
Nominal group technique	X	X	X		
SCAMPER			X		
Creative problem solving	X	X	X	X	
Six hats and lateral thinking	X	X	X		
Stakeholder analysis	X	X			
SWOT			X		
Environmental scanning			X		
Benchmarking		X			
Gap analysis			X		
Poka-yoke			X		
Quality function deployment			X		
TRIZ			X		
Y2X		X	X		
Taguchi methods			X	X	
SMED			X		
Process redesign			X		
Process capability analysis			X	X	
Control charts			X	X	
Weighted score method				X	
Decision trees				X	
Solution selection matrix				X	
Risk assessment				X	
A simple step procedure for implementation					X
Gantt charts					X
Key performance indicators		X			X
Radar diagrams		X			
Balanced scorecard		X			

REFERENCES

Ackoff, R. L. (1991) 'Continuous improvement-II', *Systems Practice*, Vol. 4, pp. 541–4.

Ackoff, R. L. (1993) 'Benchmarking', *Systems Practice*, Vol. 6, no. 6, p. 581.

Ajimal, K. S. (1985) 'Force field analysis: a framework for strategic thinking', *Long Range Planning*, Vol. 18, no. 5, pp. 55–60.

Allen, E. T. et al. (1984) *The Road of Reliability*, Motor Trade Joint Research Committee, Reliability Project Group.

American Society for Quality (n.d.) *Learn about Quality: Quality Tools*, ASQ, http://www.asq.org/learn-about-quality/new-management-planning-tools/overview/process-decision-program-chart.html (accessed 19 April 2007).

Archibald, R. D. (1992) *Managing High-Technology Programs and Projects*, Chichester, Wiley.

Bandurek, G. R., Disney, J. and Bendell, A. (1988) 'Application of Taguchi methods to surface mount processes', *Quality and Reliability Engineering International*, Vol. 4, pp. 171–81.

Bendell, A., Disney, J. and Pridmore, W. A. (1989) (eds) *Taguchi Methods: Applications in World Industry*, Kempston, IFS Publications.

Berry, L. L., Zeithaml, V. A. and Parasuraman, A. (1985) 'Quality counts in services, too', *Business Horizons*, May–June, pp. 44–52.

Boden, M. A. (1994) *La MenteCreativa, Mitos y Mecanismos*, Barcelona, Gedisa.

Booker, J. D., Raines, M. and Swift, K. G. (2001) *Designing Capable and Reliable Products*, Oxford, Butterworth Heinemann.

Bossert, J. L. (1991) *Quality Function Deployment*, ASQC Quality Press.

BS EN 60812:2006 *Analysis Techniques for System Reliability – Procedure for Failure Mode and Effects Analysis*, London, British Standards Institution.

BS IEC 62198:2001 *Project Risk Management. Application Guidelines*, London, British Standards Institution.

BS ISO 11462-1:2001 *Guidelines for Implementation of Statistical Process Control (SPC). Elements of SPC*, London, British Standards Institution.

BS 5760-7:1991 *Reliability of Systems, Equipment and Components – Part 7: Guide To Fault Tree Analysis*, London, British Standards Institution.

Business Link (2007) *Grow your Business: Setting Business Targets: What is a Key Performance Indicator?*, Business Link/Cranfield University School of Management, http://www.businesslink.gov.uk/bdotg/action/

detail?r.l1=1074404796&r.l3=1074428638&r.t=RESOURCES &type=RESOURCES&itemId=1074429057&r.i=1074429096 &r.l2=1074404826&r.s=sc (accessed 25 April 2007).

Callaghan, M. (1992) 'EIS and total quality', in Holtham, C. (ed.) *Executive Information Systems and Decision Support*, London, Chapman & Hall, pp. 171–9.

Casti, J. L. (1997) *Would Be Worlds*, New York, Wiley.

Checkland, P. B. (1972) 'Towards a systems-based methodology for real world problem solving', *Journal of Systems Engineering*, vol. 3, pp. 87–116.

Checkland, P. B. (1981) *Systems Thinking, Systems Practice*, Wiley, Chichester.

Ciucci, S. P. (1988) 'Fuel systems spring lock coupling – design of experiments', in *Taguchi Methods, Quality Engineering, Executive Briefing*, Liviona, Mich., American Supplier Institute.

Cohen, L. (1988) 'Quality function deployment: an application perspective from Digital Equipment Corporation', *National Productivity Review*, Summer, pp. 197–208.

Day, R. G. (1993) *Quality Function Deployment. Linking a Company with ItsCustomers*, Milwaukee, ASQC Press.

Defra (2006) *Environmental Key Performance Indicators*, London, Department for Environment, Food and Rural Affairs.

DTI (1989) *The Quality Gurus*, London, Department of Trade and Industry.

Evans, J. R. and Lindsay, W. M. (1996) *The Management and Control of Quality*, West Publishing Company.

Federico, M. (2005) *Rath & Strong's Workout for Six Sigma Pocket Guide*, New York, McGraw-Hill.

Finney, D. J. (1945) 'Fractional replication of factorial arrangements', *Annals of Eugenics*, Vol. 12, pp. 291–301.

Fortune, J. and Peters, G. (1995) *Learning from Failure – The Systems Approach*, Chichester, Wiley.

Fox, J. (1993) *Quality through Design*, Maidenhead, McGraw-Hill.

George, M. L., Rowlands, D., Price, M. and Maxey, J. (2005) *The Lean Six Sigma Pocket Toolbook*, New York, McGraw-Hill.

Gibis, B., Artiles, J., Corabian, P., Meiesaar, K., Koppel, A., Jacobs, P., Serrano, P. and Menon, D. (2001) 'Application of strengths, weaknesses, opportunities and threats analysis in the development of a health technology assessment program', *Health Policy*, Vol. 58, no. 1, October, pp. 27–35.

Goh, T. N., Xie, M. and Xie, W. (1998) 'Prioritizing processes in initial implementation of Statistical Process Control', *IEEE Transactions on Engineering Management*, Vol. 45, no. 1, pp. 66–72.

Hammer, W. (1972) *Handbook of System and Product Safety*, Englewood Cliffs, NJ, Prentice Hall.

Hebel, M. (1999) 'World-views as the emergent property of human value systems', *Systems Research and Behavioral Science*, Vol. 16, pp. 253–61.

Hipple, J. (2005) 'The integration of TRIZ with other ideation tools and processes as well as with psychological assessment tools', *Creativity and Innovation Management*, Vol. 14, no. 1, pp. 22–33.

Huang, G. Q. (1996) 'Developing DFX tools' in Huang, G. Q. (ed.) *Design for X – concurrent engineering imperatives*, London, Chapman and Hall.

Jenkins, D. (2005) *Targetitis: The current organisational disease*, Cirencester, Management Books 2000.

Joiner, B. L. and Scholtes, P. R. (1985) *Total Quality Leadership vs. Management by Control*, Oklahoma City, Okla., Joiner Associates Inc.

Kipling, R. (1902) 'The Elephant's Child' in *Just So Stories for Little Children*, London, Macmillan.

Lewis, P. J. (1992) 'Rich picture building in the soft systems methodology', *European Journal of Information Systems*, Vol. 1, pp. 351–60.

Marconi (2000) *Six Sigma User Guide*, Coventry, Marconi.

Mason, B. and Antony, J. (2000) 'Statistical process control: an essential ingredient for improving service and manufacturing quality', *Managing Service Quality*, Vol. 10, no. 40, pp. 233–38.

MiC Quality (n.d.) *Six Sigma Glossary*, MiC Quality, http://www.micquality.com/six_sigma_glossary/team_tools.htm#13 (accessed 19 April 2007).

Mind Tools (n.d.) *Stakeholder Analysis*, Mind Tools Ltd, http://www.mindtools.com/pages/article/newPPM_07.htm (accessed 20 April, 2007).

Montgomery, D. C. (2005) *Introduction to Statistical Quality Control*, 5th international edn, New York, Wiley.

Muñoz-Seca, B. and Riverola, J. (2004) *Problem-Driven Management*, Basingstoke, Palgrave Macmillan.

Mycoted (2006) Creativity Techniques: SCAMPER, Mycoted, http://www.mycoted.com/SCAMPER (accessed 20 April 2007).

NHS Institute for Innovation and Improvement (2006) *Service Improvement Tools: Spaghetti Diagram*, NHS Institute for Innovation and Improvement, http://www.nodelaysachiever.nhs.uk/ServiceImprovement/Tools/IT215_Spaghetti_Diagram.htm (accessed 20 April 2007).

Nikkan Kogyo Shimbun Ltd (1988) *Poka-yoke: Improving Product Quality by Preventing Defects*, University Park, Ill., Productivity Press.

Oakland, J. S. (1986) *Statistical Process Control*, Oxford, Heinemann.

Ohno, T. (1988) *Toyota Production System*, Portland, Productivity Press.

O'Reilly Media (2007) *Treemap*, O'Reilly Media Inc., http://radar.oreilly.com/archives/Q406plangtree.html (accessed 20 April 2007).

Parasuraman, A., Zeithaml, V. A. and Berry, L. L. (1985) 'A conceptual model of service quality and its implications for future research', *Journal of Marketing*, Vol. 49, no. 4, pp. 41–50.

Plackett, R. L. and Burman, J. P. (1946) 'The design of optimum multifactorial experiments', *Biometrika*, Vol. 33, pp. 305–25.

Police Standards Unit (2004) *Police Performance Monitoring Report 2003/04 – Suffolk*, http://police.homeoffice.gov.uk/performance-and-measurement/performance-assessment/perform-monitors-04 (accessed 16 May 2007).

ReVelle, J. B., Moran, J. W. and Cox, C. A. (eds) (1998) *The QFD Handbook*, New York, Wiley.

Rothwell, R. and Gardiner, J. P. (1983) *Design and Economy: the role of design and innovation in the prosperity of industrial companies*, The Design Council.

Slabey, W. R. (1991) *Planning the QFD Project*, Ford Motor Co.

Slack, N., Chambers, S. and Johnston, R. (2007) *Operations Management*, 5th edn, Harlow, Prentice Hall.

Snee, R. D. and Hoerl, R. W. (2005) *Six Sigma beyond the Factory Floor*, Upper Saddle River, NJ, Pearson.

Stapenhurst, T. (2005) *Mastering Statistical Process Control*, Oxford, Elsevier Butterworth-Heinemann.

Stoneburner, G., Goguen, A. and Feringa, A. (2002) *Risk Management Guide for Information Technology Systems*, Gaithersburg, Md, NIST.

Taguchi, G. (1986) *Introduction to Quality Engineering*, Tokyo, Asian Productivity Association.

Tait, J. and Chattaway, J. (2003) *Risk and Uncertainty in Genetically Modified Crop Development: The Industry Perspective*, Innogen working paper 1, June

Terry, A. (1995) 'Service quality in the air; quality function deployment in the service sector', in *A Selection of Research Projects Entered for the 1994 Quality Award for Theses on Total Quality Management*, Brussels, European Foundation for Quality Management.

The Open University, B822 *Creativity and Idea Generation Techniques*, Technique Selector, http://students.open.ac.uk/oubs/b822/ (accessed 17 May 2007).

The Open University (2003) T837 *Systems Engineering*, Block 6, 'Tools and techniques', Part 5 'Consensual decision making', Milton Keynes, The Open University.

US Army Enterprise Solutions Competency Center, *CPI Tools: SIPOC Process Map*, US Army ESCC, http://www.army.mil/ESCC/cpi/tools_i1.htm (accessed 20 April 2007).

Vickers, G. (1981) 'Some implications of system thinking', reprinted in Open Systems Group *Systems Behaviour*, London, Harper and Row, pp. 19–25.

White, D. (1995) 'Application of Systems Thinking to Risk Management: A Review of the Literature', *Management Decision*, Emerald Group Publishing Limited

Wu, Y.-I. (1986) *Orthogonal Arrays and Linear Graphs*, Livonia, Mich., American Supplier Institute.

You, L., Johnson, M. and Wood, S. (2004) 'Exploring strategic priorities for regional research and development (R&D) in East Africa', unpublished working paper, Washington, DC, International Food Policy Research Institute, http://www.ifpri.org/themes/sakss/saksseafrica.asp (accessed 21 February 2007).

Zeithaml, V. A., Berry, L. L. and Parasuraman, A. (1988) 'Communication and control processes in the delivery of service quality', *Journal of Marketing*, Vol. 52, April, pp. 35–48.

ANSWERS TO ACTIVITIES

Activity 3.1

The diagram shows only simple causal relationships. It does not show interrelationships between causes and the ways in which causes and effects can feed back on each other and thus amplify or masks their effects on the whole.

Activity 3.2

Activity 3.3

Inputs might include:

- opportunities for improvement
- problems
- information from customers
- quality control data
- information from suppliers.

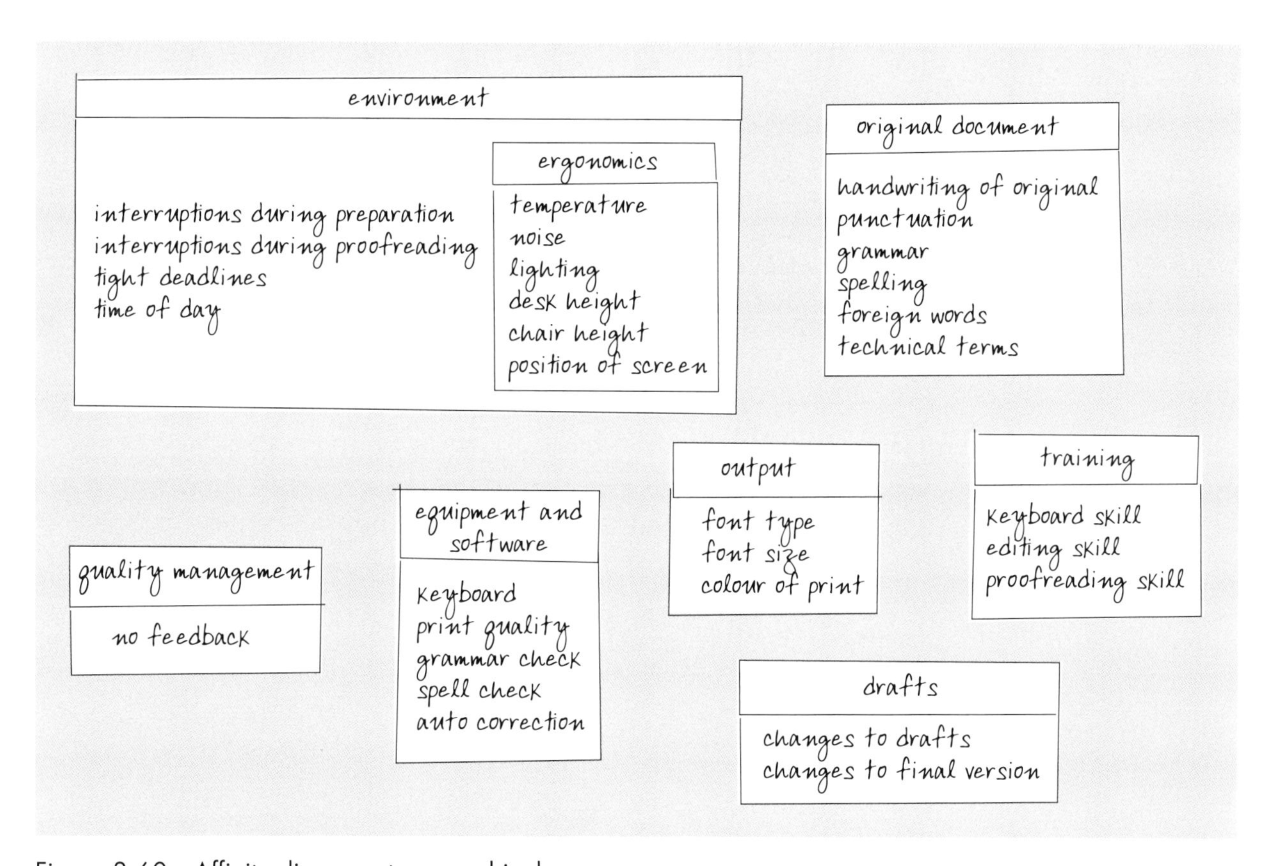

Figure 3.60 Affinity diagram: typographical errors

Outputs might include:

- suggestions for improvement
- increased performance
- greater profitability
- proposals for further quality improvement projects.

Activity 3.4

See Figure 3.61.

Activity 3.5

See Figure 3.62.

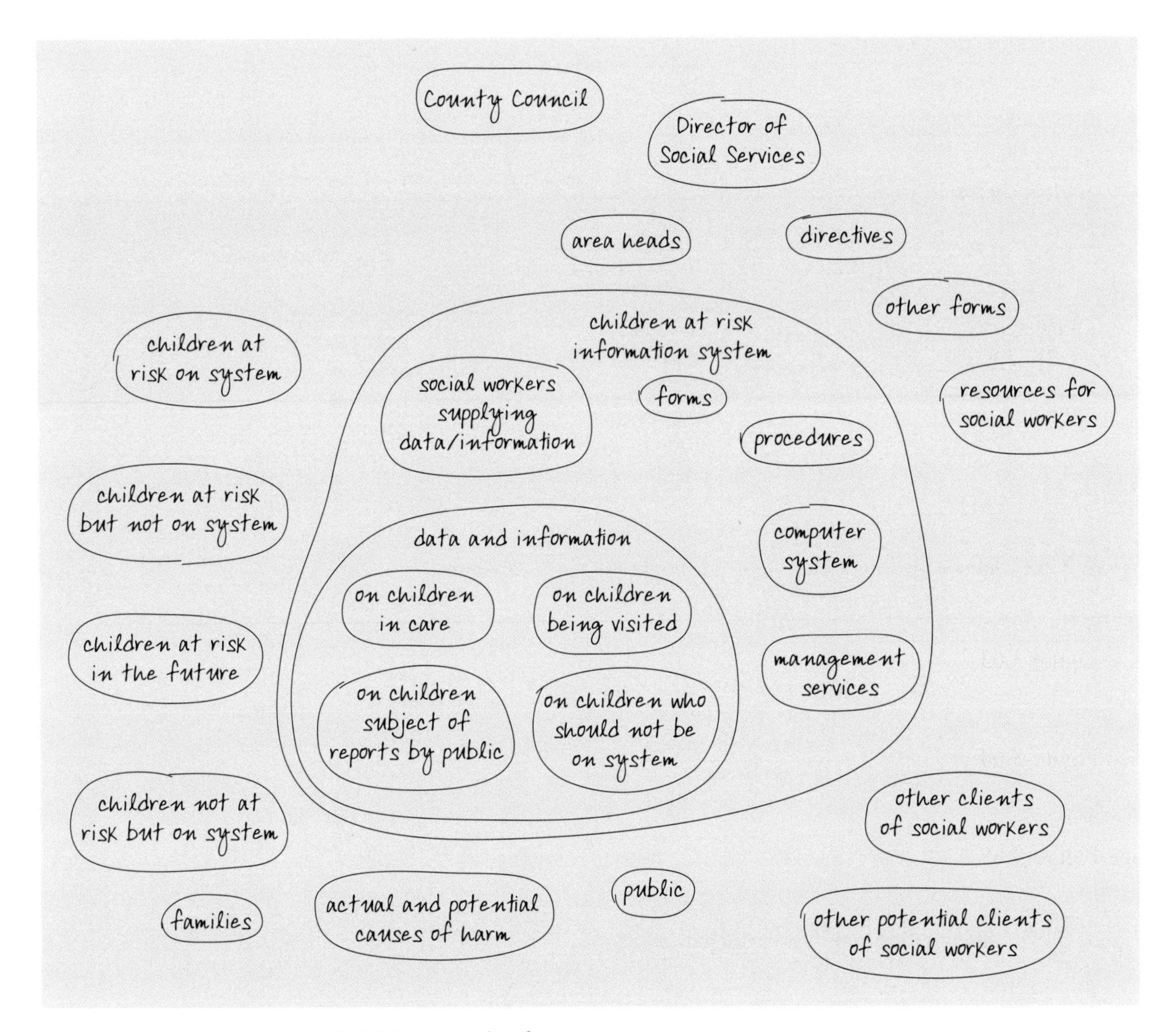

Figure 3.61 Systems map of children at risk information system

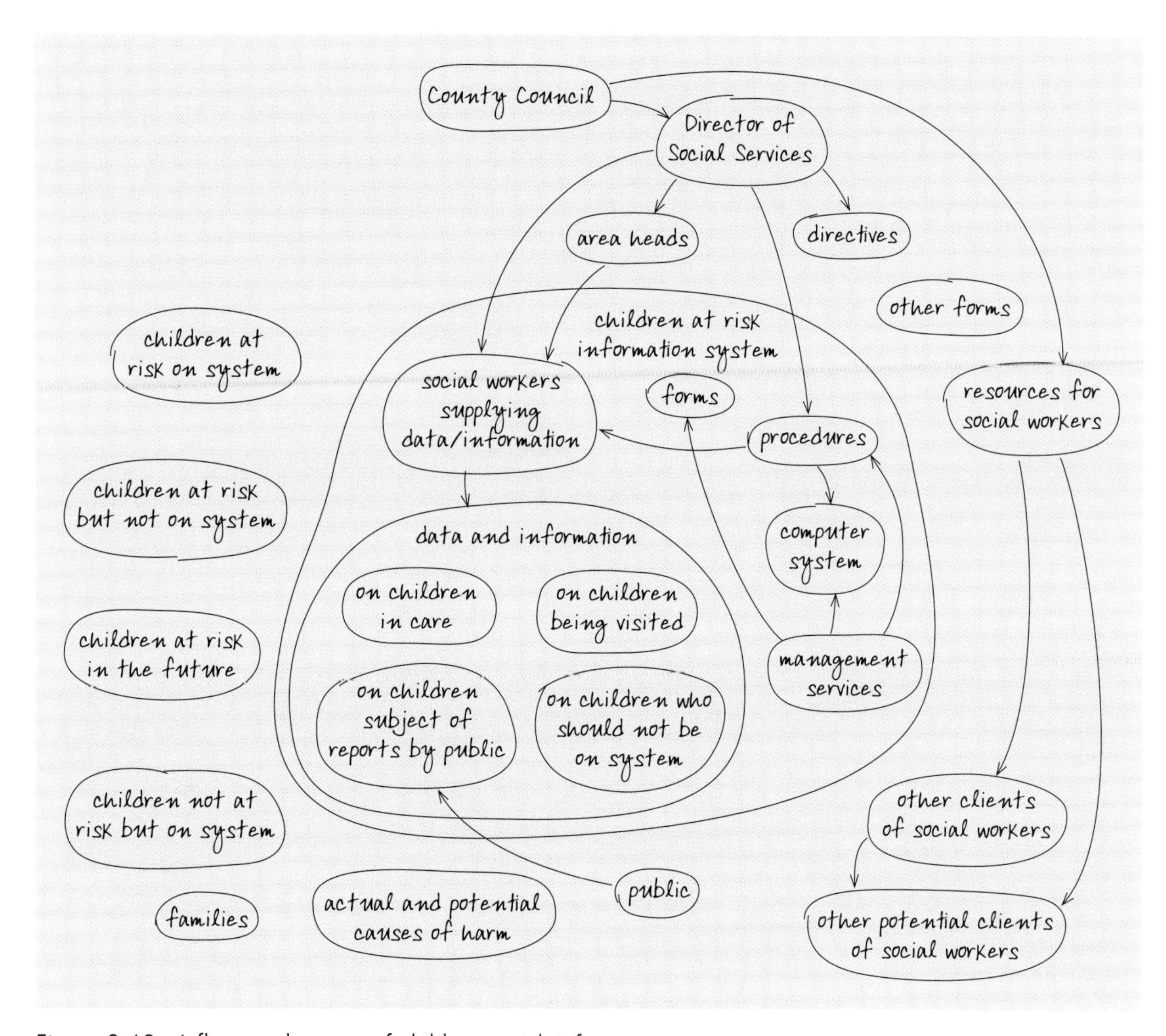

Figure 3.62 Influence diagram of children at risk information system

Activity 3.6

See Figure 3.63.

Activity 3.7

See Figure 3.64.

Activity 3.8

See Figure 3.65.

Activity 3.9

(a) Figure 3.25 has only one head: achieve environmental sustainability.
It has three tails: take organic chemicals from oil as starting material for pesticides; use energy inputs; and use new technology.

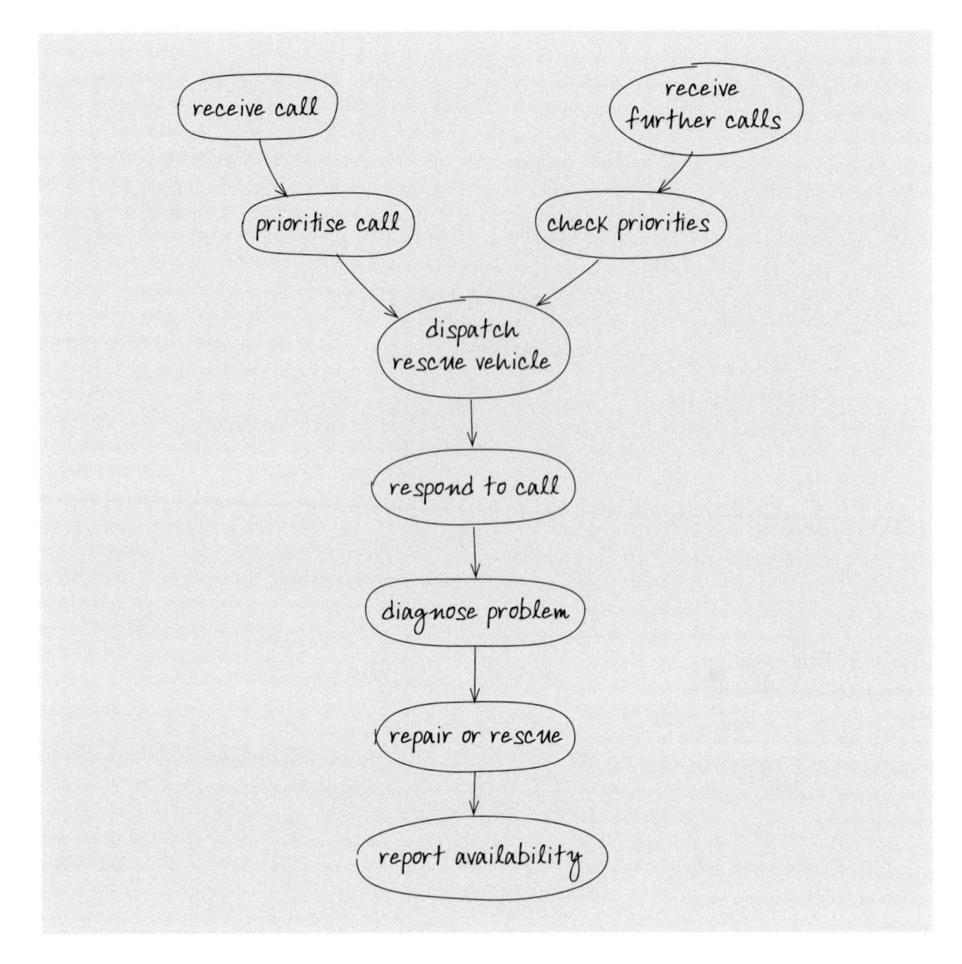

Figure 3.63 Activity sequence diagram for a car rescue service

Figure 3.26 has two heads: contribute to sustainable farming systems; and retain long-term viability for the company. It has four tails: develop new animal feed crops tailored to nutritional needs; develop 'cleaner' pesticides; develop 'functional foods'; and respond to public concerns about GM foods.

(b) My attempt to expand the construct 'use the sun's energy to give pest control... use oil and energy and produce emissions' is as follows. Using the sun's energy to give pest control, rather than using oil and energy and thus producing emissions, is a preferred approach to building environmental sustainability.

(c) Because of the negative sign on the link, I interpret the relationship between 'use less pesticide' and 'build a factory' as meaning a decision to use less pesticide will lead to not having to build factories.

Activity 3.10

Your list might include:

- brake chatter
- brake failure

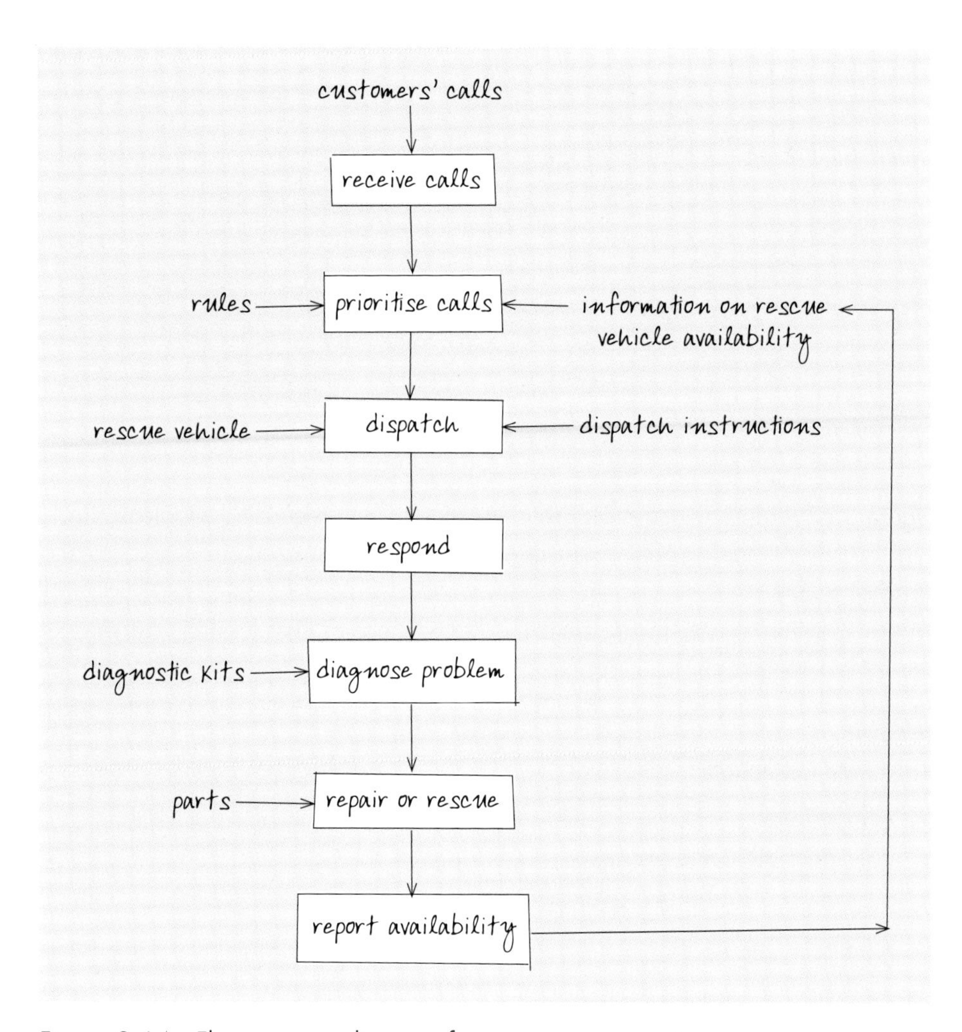

Figure 3.64 Flow-process diagram for a car rescue service

- failure of engine to start
- fuel fumes
- high oil consumption
- loss of power assist
- loss of steering
- oil leakage
- inoperative radio
- reduced vehicle performance
- surging
- warning light for oil/temperature/alternator
- water leaks
- noise.

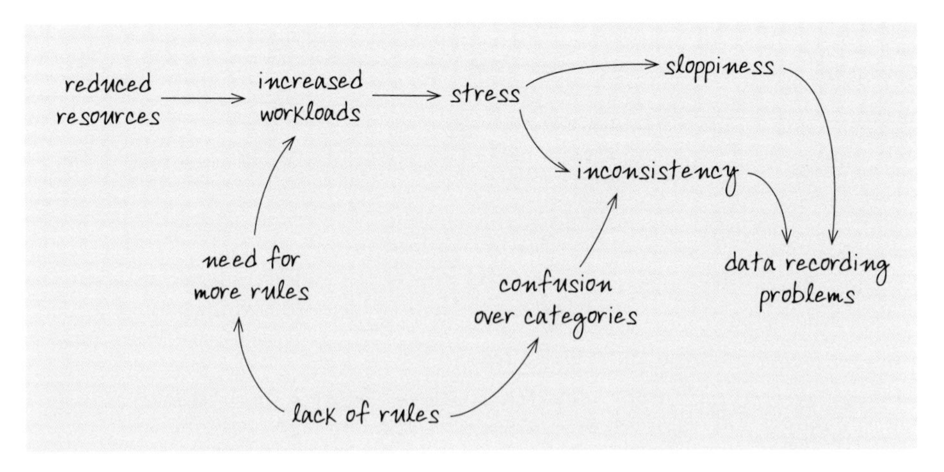

Figure 3.65 Multiple-cause diagram examining data recording problems

Activity 3.11

Rules 2, 3 and 5 will most affect the atmosphere of the session and thus its likely success.

Activity 3.12

Stakeholders include:

- competitors
- consumer advocate groups
- customers – home and overseas
- environmental groups
- EU bodies
- farm workers
- farmers
- local communities
- media
- safety and health groups
- suppliers
- trade associations
- UK government.

Activity 3.13

The environment of the system lies outside the boundary. This is not everything in the universe that is not part of the system, or something to do with the weather, but simply those things that influence the behaviour of the system or are influenced by it. The components of the environment can also exert a degree of control over the system but the environment cannot be controlled by the system.

Activity 3.14

Tangibles include:

1 notification of the appointment
2 the appearance of the service engineer
3 the attitude of the service engineer
4 punctuality
5 the service equipment.

Activity 3.15

Gap 1 is caused by lack of understanding of customers' expectations, so data gathering and review will be necessary. Techniques such as benchmarking may be helpful.

Gap 3 is caused by people failing to do what they should be doing, so empowerment and better motivation and training are indicated.

Activity 3.16

Customer requirements:

- lightweight
- ability to operate for long periods without recharging.

Design requirements:

- small batteries
- minimal power consumption while not in use.

Parts characteristics:

- low-power microphone
- low power consumption
- CMOS components.

Activity 3.17

The nearest example I can suggest is using a hand print as the equivalent of a swipe card to gain admission to a building. However, it could be argued that only the part of the mechanism that provides the signal has been replaced. The mechanism needed to read the signal and interpret it is still required.

Activity 3.18

A full factorial experiment for 20 three-level factors would require over three billion trials!

Activity 3.19

The most appropriate S/N would be type S for use when smallest is best:

$$S/N_S = -10\log_{10}\left(\frac{1}{n}\Sigma y_i^2\right)$$

This would have been selected because Bandurek et al. wished to minimise displacement.

Activity 3.22

I would use Ppk, because prioritising actions means being interested in ‘worst first’.

Activity 3.23

An organisation beginning SPC implementation may well not have well-organised historical records of defects. Without these it would have difficulty producing a check sheet.

Activity 3.24

The two difficulties I would suggest are:

1. The charts give no indication of which processes contribute to which defects. The true sources of a defect can be difficult to identify correctly.
2. Because the charts emphasise the highest counts, the organisation might focus on frequent but trivial nuisances. This might be mitigated by charting costs of defects instead of counts. This use of a Pareto chart is somewhat similar to prioritising by technical criticality, as discussed earlier.

Activity 3.25

As I interpret the standard:

- The engineering firm had the process under control and was simply steering its processes (objective b), having improved them previously.
- The police force was shown improving the performance of its process (objective c), although the need to achieve as much as possible with limited resources (the economic objective) was also mentioned.
- The neuro-physiologist sought to better understand the natural process that he was studying (objective a).

Activity 3.26

Some of the variation that is measured in a process is due to the way in which measurements have been taken, rather than due to the process itself. It may be due to use of different measuring equipment on separate occasions, perhaps different designs such as the consultant’s two types of electrode, or different instances of the same tool, or it may be due to the way the equipment is set up and used on separate occasions, whether by different operators or the same person. Indeed, several of these influences may be present together, and their effects may interact.

Activity 3.27

My impression is of much more variation between products than between operators, which implies that the measurement system is capable.

Activity 3.28

You have already seen (in the parts data) that a process can be in control but not centred within specification limits. If the reference value is the centre between the limits, a cusum would have a persistent slope. The same is true of any other arbitrarily chosen level of quality. Whoever sets such a level without being sure that the process can achieve it will be persistently reminded of failure by the cusum's slope.

In some cases the slope may be helpful. For example, the slope can communicate explicitly a discrepancy between the process and a target, a historic ideal, or an industry benchmark. Past changes in the slope would indicate how the discrepancy has changed. A slope can also indicate progress towards (or away from) a forecast value of the process mean – a horizontal line would indicate that the forecast has been reached.

Activity 3.29

In my view the genuine out-of-control conditions arising from the process rather than from data collection are:

- a temporary event such as a stand-in person
- a change in one of multiple sources of data especially in a chart of dispersion
- a sudden change in the process (e.g. damaged equipment; new staff, manager, procedure, supplier, target, bonus, etc.)
- wear and tear on machinery or measuring equipment, relaxation of procedures
- time of day, day of week, end of week/month/year effects.

Arguably the two 'tampering' causes might also be included because, although they are influenced by the data collection system, they too are faults in the way the process is being operated.

Activity 3.30

(a) The points labelled '1' are outside the control limits. However, these are not isolated points, so they signify that the process was unstable. However, stability has improved, and if another such point should occur later I think it would be more likely to be due to a temporary event.

(b) The points labelled '4' are alternating between high and low. Alternation like this normally suggests that two processes are being monitored, or that the process is being adjusted after each measurement. In this case, adjustment away from zero is very unlikely because zero is the process goal. Looking more closely at the data, there are seven, eight or nine observations per month, about two per week. The process may well be observed in alternating conditions, for example with and without weekends.

(c) I would do nothing about the type 1 points because none has occurred for six months. I would investigate the reason for alternation. If it is as I suspect, then I would look into the differences between good and bad periods. While addressing the bad periods, I might use two charts to monitor each state of the process separately.

Activity 3.31

I can think of organisations that tamper with working processes, that are enamoured of high technology, or that allow mutual mistrust to persist between managers and workforce. Any of these attitudes could create a trap like the one that ensnared Milton Technologies. However, I would be wary of assuming that such organisations could not deal with the problems. The manufacturing managers in Programme 1 recounted a story in which they met exactly such a problem and succeeded in changing their culture. The police, renowned for hierarchical organisation, spoke of creating a mindset that was subtly but profoundly different from their normal control mentality.

Activity 3.32

Because the process is in control I would use capability ratios rather than performance ratios. I would use two ratios, Cp and Cpk, because they describe different components of variation that might be addressed separately:

Cp describes the width between specification limits relative to the width of the variation. A ratio of at least 1.33 would indicate sufficient capability, unless the stricter Six Sigma capability at 2.00 is required. Cpk describes how far the process outcome is off-centre between specification limits. I would like Cpk to be as close to Cp as possible, i.e. not off-centre at all.

Activity 3.33

(a) Monitoring capability ratios is useful in any process for which a capability target has been set. It is also useful in any process where at least one specification limit falls inside control limits, because some changes will show up in adverse capability ratios before out-of-control conditions occur.

(b) I would use a performance ratio, because if a process were to become unstable I could not trust a capability ratio.

(c) Ppk shows whether a process outcome is off-centre. So, for example, a downward shift in Ppk may highlight wear in machines or changes in team working practices.

(d) Pp shows whether variation is too wide. So, for example, a downward shift in Pp may highlight a fault in a tool or a need for training.

(e) I would monitor both Ppk and Pp because together they improve diagnosis.

(f) I would look for anything that needs to be followed up. First, I would look at processes that have improvements in progress, to check whether these are taking effect. Second, I would look for shifts in capability. Shifts down may signify problems not reported elsewhere,whereas shifts up may signify lessons for other processes.

(g) I would produce I-MR charts of the capability ratios from periodic samples of the process. These would show whether the ratios are stable. I would also produce a cusum chart relative to the mean of recent samples, for early warning of shifts in capability.

(h) I could display the target as a reference line on the I chart, and perhaps also make it the reference value of a second cusum, whose varying slopes would display the history of performance relative to the target.

Activity 3.34

Starting with the furthest node:

$$\mathrm{EMV} = 0.7 \times 10 + 0.2 \times 12 + 0.1 \times 13 = 7 + 2.4 + 1.3 = 10.7$$

This value of EMV replaces this furthest node.

Going back to the next remaining node:

$$\mathrm{EMV} = 0.6 \times 10.7 + 0.4 \times 21.5 = 6.42 + 8.6 = 15.02$$

This value of EMV replaces this node.

Going back to the first node:

$$\mathrm{EMV} = 0.6 \times 15.02 + 0.4 \times 0 = 9.012$$

Activity 3.35

The three I would choose are: questions about suppliers, failure modes and effects analysis, and environmental scanning. However, there are plenty of other valid options. If your selection is different you may like to discuss this activity in the course online forum.

Activity 3.36

The answer is: not necessarily. Any delay in an activity on the critical path will delay the implementation beyond the earliest completion date but even the activities on the critical path may have float if the implementation does not have to be finished by the earliest possible date. Therefore, the target date might still be met.

APPENDIX: CREATIVITY AND IDEA GENERATION TECHNIQUES

Technique	Summary
Advantages, limits, unique qualities	To evaluate an idea from a shortlist constructively: brainstorm its advantages, then its limitations, and then its unusual qualities.
Alternative scenarios	To help medium-range planning, develop several scenarios for plausible, qualitatively distinct futures you might have to confront. Stages: state problem; identify forces; build alternative scenarios; note opportunities and synergies.
Analogies	Close analogies may help conventional understanding, but remote or bizarre analogies are often better as mental provocations Example: for challenging assumptions, idea generation, or innovation.
Analysis of interactive decision areas (AIDA)	An analytical process for exploring interactions between elements of a multi-problem situation. Stages: identify problem set, form interaction table, identify options, analyse interactions, select.
Anonymous voting	Anonymity (as in nominal groups) allows risk taking, offering some protection from interpersonal pressures. Example: facilitator displays options, members create their own shortlists privately on cards in agreed format, and then the cards are handed in face down.
Assumption surfacing	A structured process to challenge mindsets. Stages: identify stock response; identify underlying assumption; estimate its importance by potential impact of counter-assumption on stock response. Classify by importance and validity.
Attribute listing (and variants)	Basic combinatorial method for product development ideas. Stages: list attributes of existing product; identify range of options for achieving each attribute; explore different combinations of attribute options as possible new products.
Boundary examination	For problem clarification: identify key words in problem statement; examine for hidden assumptions (example: what happens if you paraphrase them?); use new understanding to redefine problem.
Boundary relaxation	Exploring an issue by examining and relaxing assumptions about its boundaries using methods such as: 'not'-ing the elements of the problem statement, seeking out further information, standard checklists, and boundary brainstorming.

Technique	Summary
Brain sketching	Pictorial variant on Pin card. Each group member privately and quickly draws a potential solution; the sketch is then passed to the person on the right for development or annotation, or to stimulate a new idea sketch; repeat until final review.
Brainwriting 635	'635' was one of the earliest brainwriting methods to be devised, and is closely related to Brainwriting pool and Pin cards. Indeed you could think of it as a tightly specified version of Pin cards. It takes 30 minutes (six rounds of five minutes) and should generate 108 ideas. Although it needs a time-keeper and organiser, it doesn't require skilled facilitation.
Brainwriting game	Idea generation game. Members write down the most implausible solutions. These are displayed and each person tries to show that others' ideas are actually plausible. Members vote for idea that remains least plausible (small prize). Then develop ideas seriously.
Brainwriting pool	Group members write ideas anonymously on slips/cards, which they put in a central pool. When they need provocation, they take a random sheet from the pool and add ideas or comments. Repeat as required for 20–30 minutes.
Browsing	Browsing through libraries, bookshops, the Web, etc. in a part-systematic and part-capricious way to gather information and ideas relevant to a particular project.
Bug listing	List many things that irritate you – sensible, crazy, silly, funny – and use them as triggers for idea generation, product development or innovation.
Bullet proofing	To check the robustness of implementation plans, brainstorm questions/challenges ('What if ...?'), and prioritise concerns in a 2×2 grid of seriousness vs. likelihood.
Bunches of bananas	Informal, provocative intervention (usually by the facilitator) using an unexpected crazy idea or action to overcome blocks in a group idea generation process. Needs sensitivity, good timing, and clear signalling.
Card storyboards	Leader puts up a problem statement card. Suggestions for solution categories are displayed as 'header' cards below the title card. Ideas are then generated and displayed under each header card (perhaps prompted by the facilitator). There are several variants.
Cartoon storyboard	Uses a six-frame storyboard that represents your future goal, your present position, and four intermediate steps to identify blocks to overcome and help strategy development, implementation planning, or issue exploration.
CATWOE	CATWOE is a mnemonic associated with the issue-defining phase of Peter Checkland's Soft systems method. However, it is also useful in its own right as a checklist of features to look for in any problem or goal definition.

Technique	Summary
Causal mapping	Causal mapping, sometimes also known as cognitive mapping, helps you create a discussable, shareable network diagram showing your beliefs about the causes and consequences of a situation.
Charrette	Work area groups (*c.* 4–12 members) meet regularly over a long period to encourage participation and communication and to improve quality, using creative problem-solving techniques. Facilities, training and management support are provided.
Clarification	Blocks can often be challenged by identifying 'fuzzy' uses of language that distort the user's self-perception, such as: deletion, unspecified verbs, nominalisations, modal operators.
Classical brainstorming	Osborn's original facilitated group process. Principles: deferred judgement, quantity breeds quality. Rules: no criticism, freewheel, quantity, hitch-hike. Group of 5–7 plus facilitator. Ideas are recorded on numbered list. Works best as part of a wider process.
ClichÕs, proverbs and maxims	This technique uses the same basic excursion process as Analogies or Random stimuli, except that it uses a clichÕ, proverb or maxim.
Collective notebook (CNB)	Each participant receives a notebook plus instructions, etc. They write one idea per day in the book for four weeks, perhaps swapping notebooks half-way, and/or being fed priming information periodically. Notebooks are then collected and processed.
Comparison tables	Construct a table with options listed vertically and criteria across the top. Rate each option against each criterion. There may be different weightings for each criterion. Sum or combine the weighted ratings for each option to give its relative value.
Component detailing	Break problem into one component per group member. Each member privately develops a detailed solution to their bit. Sub-solutions are then combined into a single overall collage 'solution', which is used as an idea-generating trigger.
Consensus mapping	To coordinate 20–30 action ideas in an overall plan: form task groups. Map clusters: first solo, then in task groups, then overall. Map action sequences: first within each task group; then presentations between task groups; then final synthesis.
Constrained brainwriting	A brainwriting process adapted so that the range of ideas generated is narrowed down to a predetermined focus, either by priming the idea flow with selected ideas or by providing thematic headings for response sheets.
Controlling imagery	This set of techniques is intended to help you to exercise more control over your imagery, both in the positive sense of doing more with it, and in the negative sense of knowing how to stop it or defuse it.

Technique	Summary
Crawford slip writing	All those attending a large group/gathering are issued with a pencil and a pack of paper slips. At key points in the presentations, they are asked to write ideas on the slips, which are collected immediately. Ideas are sorted and a report is fed back to participants.
Creative problem solving (CPS)	Classic facilitated problem-solving process. Stages: mess finding; data finding; problem finding; idea finding; solution finding; acceptance finding. Each stage has a convergent and divergent phase, and involves subsidiary techniques.
Criteria for idea-finding potential	Six criteria for evaluating problem statements: idea generation potential, focus, ownership, affirmative orientation, freedom from criteria, brevity and clarity.
Critical path diagrams	Project planning map: arrows represent activities that take time and cost money ('build wall'); nodes represent events ('bricks arrive') that start/end an arrow. The 'critical path' through this network determines the minimum duration of the project.
Decision seminar	A social policy group works over an extended period from a specially equipped room using a standardised conceptual framework based around 'five intellectual tasks', 'seven information-gathering strategies', and using a 'seven-step decision process'.
Delphi	For gathering opinions from perhaps 15–30 informed people using successive questionnaires. For example, the first poses the question; the second feeds back the answers for rating and comment; a third feeds back the ranked ratings for further comment.
Dialectical approaches	Group methods using creative conflict to challenge assumptions/perceptions and to make decisions/plans, using, for example, a critical devil's advocate, proposal/counter-proposal groups.
Dimensional analysis	This is a checklist related to Five Ws and H and of most use as an aide-memoire for preliminary exploration of a problem, or perhaps for evaluating options, particularly problems with a human-relations, rather than a technical, core.
Disney	The Disney strategy uses three modes of thinking: the dreamer, the realist and the critic. Disney made extensive use of this approach in his creative work.
Drawing	Use of expressive and imaginative drawing for: personal insight; strategy development; group communication; issue exploration. Taps intuitive metaphors, symbols, etc.
Essay writing	The simple process of writing freely can help a manager string a whole jumble of ideas together. It allows plenty of scope for imagination, speculation and creative flair.

Technique	Summary
Estimate–Discuss–Estimate	To allow knowledge of overall group preferences to inform individual judgements, gather individual judgements, pool them, discuss the pooled judgement, and then have a second round of individual decisions and pooling.
Exaggeration (magnify or minify)	Magnify (or stretch) and minify (or compress) are two of the idea-generating transformations in Osborn's original checklist. They are both forms of exaggeration.
Excursions	A mental or physical 'tour', that takes your attention away temporarily from the issue you have been focusing on, can allow or provoke new ideas about it. These link to the issue metaphorically.
Factors in 'selling' ideas	When promoting an option to management – or any audience – consider both the context and the selling content.
Fishbone diagram	'Fishbone'-shaped diagram ('fish-head' = problem, 'bones'/spurs = causes and sub-causes) that maps and explores issues and interactions, prompting group discussion and reflection.
Five Ws and H	Checklist of basic questions: Who? Why? What? Where? When? How? Also known as 'Kipling's list' or 'Kids' kit', and the basis for various other checklists. Useful for all stages of problem solving.
Flow charts for action planning	Conventional flow charts can be used to represent action plans, especially where the flow will vary depending on the decisions made. They show sequences, activities, decision points, continuation logic, start points, end points and temporary exits.
Focus groups	Issue-focused sessions in which experts are invited to describe opportunities or provide information to policymakers. Example: for product marketing, development.
Focusing	The focusing technique is a form of imagery work based on body feelings and sensations rather than visual imagery.
Force field analysis	A situation analysis diagram in which driving and restraining forces are depicted as pressing the status quo towards or away from your goal. Arrows usually show the relative strengths of the different forces.
Force-fit game	Idea generation game. Groups A and B (2–8 in each) agree on a problem. Group A suggests an idea remote from problem; B has two minutes to try to develop a plausible solution. Referee gives point to B if plausible, to A otherwise. Swap and repeat as you wish.
Free association	This is an element of most other idea-generating methods, and relies on the mind's 'stream of consciousness' and network of associations.

Technique	Summary
'Fresh eye' and networking	Other people are a rich source of ideas (or of idea triggers) and information, and are often enablers and facilitators of action. Networking refers to communication to and from other people – making contacts.
Gallery method	A brainwriting method that uses flip-chart sheets placed all round the room. Participants choose a sheet and write ideas privately on it. After a while, they look over others' sheets, then each participant returns to their own sheet and adds further ideas. Repeat as you wish.
Gap analysis	Systematic, analytical search of the major dimensions of a technology for regions not currently filled by any product, since these may offer opportunities for speculation, innovation, or product development.
Goal orientation	This is a simple rational checklist for problem statements. The CATWOE checklist contains a more elaborate set of rational criteria.
Greetings cards	Group members create collage 'greetings' cards before the problem is presented. The problem is then presented and the cards are used as stimuli to generate problem-solving ideas.
Heuristic ideation technique (HIT)	A variant on Attribute listing for new product work. Stages: select two different products, A and B; list attributes of A along top of a table and attributes of B down the side; develop product ideas from table cells representing interesting combinations.
Highlighting	Given a large list of items, select any intriguing ones and sort them into clusters to identify 'hotspots'; identify what each hotspot means to you and select the hotspot that best satisfies your requirements.
Idea advocate	From a shortlist of 3–6 options, each option is allocated to someone as its 'idea advocate'. A period for research, etc. may be allowed. Each idea advocate then presents a case for their idea to the decision makers and the other idea advocates.
Imagery for answering questions	A guided imagery technique. Formulate your question clearly, relax, remember the question. Visualisation script: boat journey, lake, underground passage, emerge to brilliant sunshine, receive message, return. Write down what happened.
Imagery manipulation	Problem solving by interactive work with imagery. Needs a skilled helper. Stages: privately identify 5–7 key problem elements and devise symbols for each; then describe symbols; jointly explore and expand imagery; reach resolution.
Implementation checklists	Checklist of common implementation blocks: resources, motivation, resistance, procedures, structures, policies, risk, power, clashes, climate.

Technique	Summary
Improved nominal group technique	A facilitated, structured, group process allowing anonymous, written, idea generation and selection. Developed from the Nominal group technique, with an added pre-meeting stage, then idea collection, display, serial discussion, and voting.
In-and-out listening	Let your attention drift in and out of focus on the issue itself, note any associations for use as springboards to prompt ideas.
Interpretive structural modelling (ISM)	'Paired-comparison' method (usually computer managed) for exploring multi-factor issues (e.g. public sector prioritisation). Stages: list 10–50 items (e.g. options, objectives); compare/order every pair; map onto network diagram.
Itemised response	This approach itemises first an idea's good points, then its drawbacks, and then attempts to counter the drawbacks and transform the idea into a useful one.
Kepner Tregoe	Rational, structured process. Problem analysis: observe deviation; prioritise; explore; what is/is not; potential and possible causes; test. Decision making: requirements; priorities; options; evaluate; select; adverse consequences; implementation plan.
KJ method	Kawakita's method of clustering and diagramming a collection of elements of an issue is similar to the Snowball technique, but with a meditative element.
Laddering	Ladder up to categories that a given idea belongs to, or ladder down to sub-categories of the idea. By laddering up and down alternately you can move into new idea domains and explore broad or specific elements/areas/ issues.
Less competitive methods of voting	It may be safer to use voting to eliminate weak items rather than to select strong ones, because that way you are less likely to reject ideas that seem inappropriate at first sight, but are actually valuable.
Listing	A variant on Attribute listing for new product work. List 10–12 existing products in a relevant area (e.g. kitchen products) along both axes of a matrix. Develop product ideas from interesting combinations (e.g. toaster/kettle).
Listing pros and cons	Generate realistic list of advantages/disadvantages for each option to be evaluated. For a short list this may be enough on which to base a choice. For a longer list, merge all to form an overall prioritised set of criteria to be applied systematically to every option.
McKim's method	McKim draws on creative strategies in visual thinking to solve problems.

Technique	Summary
Metaplan information market	Uses a carefully planned and facilitated in-house interactive 'exhibition' of issues and ideas, usually around a group of strategic concerns. It is designed to generate active participation, networking, communication, consultation, polling and discussion.
Mind-mapping	The term 'Mind-mapping' was devised by Tony Buzan for the representation of ideas, notes, information, etc. in radial tree diagrams – sometimes also called 'spider diagrams'. These are now very widely used. Try a Web search on 'Buzan', 'mind-map' or 'concept map'.
Morphological analysis	Modern extension of Attribute listing. List every element of a product or process and all forms each element could take. Explore all possible combinations. For large numbers of elements and forms, simplifying algorithms and computer support are essential.
Multiple redefinition	A set of six provocative reframes for solo use: to help you break out of mindsets; to prompt issue exploration, assumption challenging and wishing; and to tap into the subconscious.
Negative brainstorming	Brainstorming ideas that criticise existing ideas, in order to identify weaknesses. Needs to be used with care because of its negativity. Normally best combined with constructive solving of the problems identified.
Nominal group technique (NGT)	NGT is a structured form of brainstorming or brainwriting for a group with a skilled facilitator and 5–9 members (or, in a suitable room, up to about 3–4 groups of 5–9 in parallel, with a separate recorder for each and a single facilitator overall).
Nominal-interacting technique	To increase mutual understanding and reduce tensions, informal 'anteroom sessions' are introduced at key points in a brainwriting process as 'refreshment breaks' during which members are encouraged to lobby, bargain, exchange information, etc.
Notebook	If you have to work on your own, you can achieve some of the benefits of brainstorming, brainwriting, etc. if you trade time for group stimulation.
Osborn's checklist	Substitute, Combine, Adapt, Modify, Magnify, Minify, Put to other uses, Eliminate, Rearrange or Reverse something in the problem is a classic creativity checklist useful for developing earlier ideas produced during brainstorming.
Other people's definitions	This is a simple approach to understanding a problem by allowing other people, with different perspectives, to challenge your view of it.
Other people's viewpoints	The real 'last word' is that of the organisation or any people whose consent and compliance are needed if anything concrete is to happen. It is vital to understand their viewpoints.

Technique	Summary
Paired comparison	To prioritise a set of fewer than 12 options, etc., set up a matrix with one cell per possible pairing. In each cell enter the more important item and a rating. Sum ratings for each option over the matrix. For more than 12 items, use Interpretive structural modelling.
Panel consensus	Experts generate 400–500 ideas; these are divided between 15 panels of 15; their shortlists are combined and sent in parallel to three panels of 5; their shortlists are combined for a top panel of 5; their shortlist goes to the final decision-making panel.
Paraphrasing key words	Two closely related techniques: replacing key words with synonyms, and using synonym pairs to trigger ideas.
Personal balance sheet	Uses a 2×2×2 table to record tangible/subjective gains/losses for self/others for each option to be evaluated. Mainly for evaluating personal options; commitment is often more effective if the table is completed in the presence of, say, a counsellor.
Phases of integrated problem solving (PIPS)	Phases of integrated problem solving is a variant of classic Creative problem solving that also defines the interpersonal steps needed.
Phillips 66 ('Buzz' sessions)	A large group (20–100) splits into 6–12 break-out groups, to spend 6–30 minutes discussing/exploring the issue, generating ideas, then reporting back.
Pictures as idea triggers	Various authors have described forms of brainstorming, brainwriting or excursion in which pictures are used as idea triggers (e.g. Batelle-Bildmappen-Brainwriting, Visual Synectics). Sometimes the pictures are first created by the group (e.g. see Greetings cards and Component detailing).
Pin cards	Group members sit round a table. Each writes ideas on cards or Post-its. Each completed card or Post-it is placed in a pile available to the person on the right. The pile from the person on the left is available as triggers to generate ideas.
Pluses, potentials and concerns	To evaluate an idea from a shortlist constructively: brainstorm its advantages, then its limitations, and then its unusual qualities.
Potential-problem analysis (PPA)	Rational, analytical evaluation of options: define key activities; list and explore potential problems; list potential causes of these problems and estimate their likelihood; develop preventive measures where possible and contingency plans otherwise.
Preliminary questions	A checklist of questions (each beginning with one of: who, what, where, why, when) used as a quick, semi-structured way to explore issues, challenge assumptions, provoke ideas, consider stakeholders, and reframe perceptions.

Technique	Summary
Problem inventory analysis (PIA)	Develop a list of generic consumer problems. Use it as the basis for a questionnaire that lists the problems but leaves the product names blank ('....... never pours properly'). Use survey answers to identify potential areas of demand.
Problem-centred leadership (PCL)	This technique describes what the leader (or facilitator) should be doing at different stages of a creative problem-solving process, both procedurally, and to support a healthy group process.
Progressive hurdles	Formal institutional idea-screening procedure by series of evaluative hurdles: culling hurdle (low-cost criteria), rating hurdle (medium cost), scoring hurdle (higher cost), in-depth analysis (complex, high-cost investigation). Few options reach the last stage.
Progressive revelation	To discourage premature closure, the problem is presented initially in very abstract form, and then in progressively more concrete forms: present abstractly; group brainstorms; re-present less abstractly; more brainstorming; etc.
Q-sort	Process for efficient ranking/prioritising lists of 60–90 items: calculate a set of class bands assuming a normal distribution over categories; then select items for the top and bottom classes. Repeat for each successive outermost class.
Quality circles	Work area groups (*c.* 4–12 members) meet regularly over a long period to encourage participation and communication and to improve quality, using creative problem-solving techniques. Facilities, training and management support are provided.
Quota	Setting an artificial quota – a pre-determined number of fundamentally different ideas that must be generated – can help you to come up with new approaches.
Random stimuli of various kinds	Use of any random stimulus or new experience, either deliberately selected (e.g. opening a dictionary at random) or opportunistic (e.g. openness to things that happen to you) to generate ideas.
Receptivity to ideas	Enhancing receptivity to ideas by use of informal communication techniques such as checking understanding by paraphrasing, or using a developmental response to build on what is valuable in the idea rather than picking on its weaknesses.
Reframing values	An individual or group process to identify and challenge values and assumptions by using the reversal of bipolar concepts to reframe perceptions and their implications.
Relational words	List of 57 relational words (about, during, opposite, while, etc.). Get a new perspective by adding a relational word to any idea or take two ideas and join them with a relational word.
Relaxation	Relaxing, with the aid of spoken instructions or suggestions, to help overcome blocks.

Technique	Summary
Reversals	Challenge mindsets by reversing some aspect of a problem (its perspective, its values, word order, phase, roles, flow, direction, stereotypes, etc.). Sometimes this may generate ideas directly, but often it is better to then re-reverse.
Rich pictures	Rich pictures are the starting point of the Soft systems method which was developed by Peter Checkland at the University of Lancaster and further adapted by the Systems Group of The Open University.
Rolestorming	Brainstorming, with group members contributing as though they were someone else – someone you know or someone in public life – in order to reduce inhibitions and switch perspective. See also Super heroes.
Sculptures	The group creates a sculpture (from a range of easily available materials/objects) that somehow evokes the problem. This is used to stimulate idea generation.
Search conference	A conference-like process to help stakeholders explore the forces that have shaped and may shape their organisation, by examining social trends, environmental change, the forces and constraints they must work under, and possible actions.
Semantic intuition	Identify problem area (e.g. 'gardening'). Choose a grammatical word-pair structure (e.g. noun-verb). List problem-related words of each type (e.g. gardening nouns and verbs). Triggers ideas with word pairs from lists (e.g. 'gnome-fertilising').
Sequential-attributes matrix	Apply a checklist such as SCAMPER to each stage in turn of any sequential procedure, noting that each stage must still lead correctly into the next, however it is changed. Best done by creating a matrix of stages vs. checklist questions.
Simple rating methods	Prior to evaluating and selecting items (e.g. potential solutions, options) from a collection, it may be helpful to rate the items according to relevant criteria or arrange them in rank order for ease of comparison.
Sleight of mouth	Offers various ways of reframing a point of view.
Snowball technique	Write each idea (or issue/action/stakeholder, etc.) on a separate card or Post-it and then cluster them into thematic groups. Can be done solo or in small groups. See also KJ method.
Soft systems method	The staged problem-solving approach developed by Checkland and adapted by the OU Systems group.
Stakeholder analysis	Identify stakeholders and table their high/low power and high/low likelihood of impact in a 2×2 matrix. Stakeholders in each of the four cells probably need to be handled differently, e.g. high power/high impact stakeholders clearly need careful management.
Sticking dots	Given a list of *N* ideas, each group member is issued with a number (perhaps 5–10% of *N*) of sticky dots, or a ration of 'ticks', and then allocates these as votes within the list. They can give more than one 'dot' to one item if they wish.

Technique	Summary
Stimulus analysis	Obtain 10 very different objects that are unconnected to the problem. Select one and list its characteristics in detail. Use each characteristic to generate ideas. Repeat with the other nine objects.
Story writing	Explore a situation and generate ideas by drawing, writing or telling a subjective story that parallels the situation in some way but is not a direct description of it. Include emotions, values, expectations, etc. See also Drawing.
Strategic assumption testing	A facilitated group process for exploring assumptions about stakeholders and preferences (e.g. in public or voluntary sector). Stages: identify clients; form groups; discuss within groups; surface, test and rank assumptions; discuss between groups.
Strategic choice approach	A facilitated manual or computer-based approach using AIDA to resolve complex problems where the range of interacting components leads to very many possible combinations of actions (e.g. many public sector strategy issues).
Strategic management process	A facilitated group process for medium-scale strategic or multiple stakeholder contexts (e.g. public or voluntary sector). Stages: historical and situational assessments; strategic issue agenda; possible options; feasibility; implementation.
Strategic options development and analysis (SODA)	The SODA method allows people to externalise their different understandings of a situation, and to develop an explicit, practical, understanding that they can share.
Successive element integration	Two people read out an idea each; the group combines these to make a third idea. Another person reads out an idea; this is combined with the previous combination. Continue until all ideas are combined. Final and interim ideas are all valid ideas.
Super group	Combines Creative problem solving with Focus groups. Used in new product development.
Super heroes	Rolestorming in which group members generate ideas based on the viewpoints and/or character attributes of chosen (super)heroic characters (Superman, Dr Who, James Bond, etc.).
Synectics	A classic structured process: problem-owner headline; group generate springboards (wishing, reframing); owner selects interesting springboard; excursions to generate solution ideas; itemised response to develop ideas; recycle as you wish.

Technique	Summary
Systematised direct induction	A group of 4–100 people sit in fours to write ideas on paper slips. After a practice exercise, participants write 'How to ...' with respect to the issue. After discussion, further problems and solutions are generated and all slips are collected and analysed.
Technology monitoring	Ongoing monitoring of the organisation's environment for opportunities for technology advances or product development, e.g. by networking, information gathering and recording, active search, speculation, etc.
Think tanks	A well-established method for providing a stimulating environment for interdisciplinary communication, networking among experts, data collection, and consultation. Often used in the public sector to deal with large-scale, long-term, value-laden issues.
Timeline	A way of using movement around a line to help imagine what a situation was/is/will be.
Transactional planning	A process for managing effective communication between planners and sponsors in political, strategic or multiple stakeholder contexts, e.g. the public or voluntary sector. Stages: formulation; conceptualisation; detailing; evaluation; implementation.
Trigger method	Using any brainstorming variant, get to the point where participants have generated a collection of ideas. Then select a batch of 5–10 ideas (e.g. one round robin or 5–10 Post-its) and use these as a focused trigger to develop further ideas. Repeat as you wish.
Using 'crazy' ideas	Crazy 'get fired' ideas can trigger idea generation (by Free association, Force fit, Excursion, or analysing strengths and difficulties) and can revive imagination and energy in a stuck or flagging group. See also Bunches of bananas.
Using experts	Using expert-to-expert sessions, Delphi-type surveys, etc. to access and collect expertise and information to support opportunity research, issue exploration, product development, etc.
Value brainstorming	An individual or group process, using brainstorming and ranking, to identify public and hidden values, and explore their implications for action.
Value engineering	Value engineering function analysis is similar to problem finding in Creative problem solving.
Visual brainstorming	Recording generated ideas as a series of quick sketches. Display them on a wall or large area, and then evaluate them using both right and left sides of the brain, e.g. view imaginatively, compare, regroup, write notes.
Visualising a goal	This is an affirmation method for focusing your energy and motivation on a particular goal. Stages: set goal; create clear mental image; focus on it often; affirm; continue as appropriate; acknowledge achievement.

Technique	Summary
Well-formed outcome	Positive and precise way of defining what you want.
Who are you?	Two ways of exploring personal identity: (1) repeatedly ask, 'Who are you?' (solo or partnered version); or (2) use guided imagery (relax; imagine being 5, 12, 25, 40, 65, older, dying, rebirth; return; reflect).
'Why?' etc. – repeatable questions	Questions you can repeat to get information can uncover unexplored factors. They can also generate answers that are often suitable for diagramming, and are very 'portable' and easy to use unobtrusively.
Wishing	Wish positively, ambitiously and imaginatively, e.g. by visualising a future, articulating desires, or imagining a path to where you want to go.
Working with dreams and images	One way of working with dream material. Stages: enter dream; study it; become images; integrate viewpoints; rework and develop; appreciate and reflect.

(Source: adapted from B822, Technique Selector)

ACKNOWLEDGEMENTS

Grateful acknowledgement is made to the following sources:

Tables

Table 3.4: Booker J.D., Raines M., Swift K.G., (2001) *'Techniques for Redesigning Products and Processes',* Designing Capable and Reliable Products, Butterworth-Heinemann Limited; Table 3.5: Fox J., (1993), *'Quality through Design'*, McGraw-Hill Book Company; Table 3.6: Snee R.D., and Hoerl R.W., (2005) *'Six Sigma Beyond the Factory Floor'*, Pearson Education Limited; Table 3.7: *'Illustrative Example of a Severity Classification for End Effects'*, 'Analysis Techniques for System Reliability – Procedure for Failure Mode and Effects Analysis (FMEA), BS EN 60812, (2006), British Standards Institute; Table 3.8: Hammer W., (1972), '*Detailed Analyses Prototypes*', Handbook of System and Product Safety, Prentice Hall, Inc; Table 3.20: George M.L., Rowlands D., Price M., and Maxey J., ' *The Lean Six Sigma Pocket Toolbook'*, McGraw-Hill Book Company; Table 3.21: Stoneburner G., Goguen A., Feringa A., (2007), *'Sample Safeguard Implementation Plan Summary Table'*, Information Technology Laboratory, National Institute of Standards and Technology; Table 3.22: Department of the Environment, (2006), *'Environmental Key Performance Indicators: Reporting Guidelines for UK Business'*, Crown copyright material is reproduced under Class Licence Number C01W0000065 with the permission of the Controller of HMSO and the Queen's Printer for Scotland.

Figures

Figure 3.2: Sixsigma. (2006), *'The General Tool Kit'*, Marconi Communications plc; Figure 3.4: Evans J.R., and Lindsay W.M. (1996), *'The Management and Control of Quality'*, by Permission of South-Western College Publishing a Division of International Thomson Publishing In., Cincinnati Ohio 45227; Figure 3.3: Evans J.R., and Lindsay W.M. (1996), '*The Management and Control of Quality*', by Permission of South-Western College Publishing a Division of International Thomson Publishing In., Cincinnati Ohio 45227; Figure 3.9: Kogyo N., (1988), *'Poka-Yoke: Improving Product Quality by Preventing Defects'*, Productivity Press; Figure 3.10: © 1998-2007 MIC Quality; Figure 3.11: Taque N.R., (2004), *'The Quality Toolbox'*, ASQC Quality Press; Figures 3.12 and 3.13: Ciucci S.P., (1988), *'Taguchi Methods, Quality Engineering Executive Briefing',* American Supplier Institute; Figure 3.14: Slack N., Chambers S., Johnston R., (2007), *'Operations Network for a Plastic Homeware Company & Shopping Mall',* Operations Management, Pearson Education Limited;

Figure 3.20: Source NHS for Innovation and Improvement- www.institute.nhs/NoDelaysAchiever; Figure 3.21: *'SIPOC Process Map Example',* Enterprise Solutions Competency Centre, Software Engineering Centre, Belvoir; Figure 3.23: Ajimal K.S., (1985), 'Force Field Analysis: a Framework for Strategic Thinking', Long Range Planning; Figure 3.25: Tait J., and Chataway J., (2003), 'Uncertainty in Genetically Modified Crop Development', The Industry Perspective, Innogen Research Centre; Figure 3.26: Tait J., and Chataway J., (2003), 'Company Perspective on Sustainable Development and Commercial Viability', The Industry Perspective, Innogen Research Centre; Figure 3.27: Allen E.T., et al, (1984), 'The Road of Reliability', Motor Trade Joint Research Committee; Figure 3.28: Hipple J., (2005), 'The Integration of TRIZ with other Tools and Processes as well as with Psychological Assessment Tools', Blackwell Publishing; Figure 3.29: 'Power/Interest Grid for Stakeholder Prioritisation', (1995-2005), © Mind Tools; Figure 3.30: Gibis B., Artiles J., Corabian Paula. Et al (2001) 'Application of Strengths, Weaknesses, Opportunities and Threats Analysis in the Development of a Health Technology Assessment Program', Elsevier Science Ireland Limited; Figure 3.32: 'Treemap', http://radar.oreilly.com/archives/Q406plangtree.html, O'Reilly Media; Figure 3.33: Parasuraman Z., Zeithaml V.A., and Berry L.L., (1985), 'A Conceptual Model of Service Quality and its Implications for Future Research', American Marketing Association; Figures 3.34 and 3.35: Shimbun N.K., (1988), 'A Poka-Yoke Exercise on Speaker Box Assembly' Productivity Press; Figure 3.37: Slabey W.R... (1983), 'Planning the QFD Project', Ford Motor Company; Figure 3.40: Day R.R., (1993), 'Quality Function Deployment Linking a company with its Customers', ASQC Quality Press; Figure 3.42: Sixsigma. (2006), 'The General Tool Kit', Marconi Communications plc; Figures 3.43, 3.44, 3.45 and 3.46, Bandurek G.R., Disney J., and Bendell. (1988), 'Quality and Reliability Engineering International', John Wiley & Sons Limited; Figure 3.47: Goh T.N., Xie M., (1998), 'Prioritizing Processes in Initial Implementation of Statistical Process Control', IEEE Computer Science Publications; Figure 3.48: Portions of the input and output contained in this publication are printed with the Permission of Minitab Inc. Minitab® and all other trademarks and logos for the Company's products and services is the exclusive property of Minitab Inc. All other marks referenced remain the property of their respective owners. See minitab.com for more information; Figure 3.49: Montgomery D.C., (2005), 'Introduction to Statistical Quality Control', John.Wiley & Sons; Figure 3.55: White D., (1995), 'Application of Systems Thinking to Risk Management: A Review of the Literature', Management Decision, Emerald Group Publishing Limited; Figure 3.57: Archibald R.D., (1992), 'Part II Managing Specific Projects, Chapter 10 Planning Projects' Managing High-Technology Programs and Projects, John Wiley & Sons Inc.; Figure 3.58: '*Police Performance Monitoring 2003/04*', Crown copyright material is reproduced

The paper used in this publication contains pulp sourced from forests independently certified to the Forest Stewardship Council®(FSC®) principles and criteria. Chain of custody certification allows the pulp from these forests to be tracked to the end use (see www.fsc-uk.org).